The Information in Contingency Tables

STATISTICS: Textbooks and Monographs

A SERIES EDITED BY

Volume 1: The Generalized Jackknife Statistic, *H. L. Gray and W. R. Schucany*

Volume 2: Multivariate Analysis, *Anant M. Kshirsagar*

Volume 3: Statistics and Society, *Walter T. Federer*

Volume 4: Multivariate Analysis: A Selected and Abstracted Bibliography, 1957-1972, *Kocherlakota Subrahmaniam and Kathleen Subrahmaniam* (out of print)

Volume 5: Design of Experiments: A Realistic Approach, *Virgil L. Anderson and Robert A. McLean*

Volume 6: Statistical and Mathematical Aspects of Pollution Problems, *John W. Pratt*

Volume 7: Introduction to Probability and Statistics (in two parts) Part I: Probability; Part II: Statistics, *Narayan C. Giri*

Volume 8: Statistical Theory of the Analysis of Experimental Designs, *J. Ogawa*

Volume 9: Statistical Techniques in Simulation (in two parts), *Jack P. C. Kleijnen*

Volume 10: Data Quality Control and Editing, *Joseph I. Naus*

Volume 11: Cost of Living Index Numbers: Practice, Precision, and Theory, *Kali S. Banerjee*

Volume 12: Weighing Designs: For Chemistry, Medicine, Economics, Operations Research, Statistics, *Kali S. Banerjee*

Volume 13: The Search for Oil: Some Statistical Methods and Techniques, *edited by D. B. Owen*

Volume 14: Sample Size Choice: Charts for Experiments with Linear Models, *Robert E. Odeh and Martin Fox*

Volume 15: Statistical Methods for Engineers and Scientists, *Robert M. Bethea, Benjamin S. Duran, and Thomas L. Boullion*

Volume 16: Statistical Quality Control Methods, *Irving W. Burr*

Volume 17: On the History of Statistics and Probability, *edited by D. B. Owen*

Volume 18: Econometrics, *Peter Schmidt*

Volume 19: Sufficient Statistics: Selected Contributions, *Vasant S. Huzurbazar (edited by Anant M. Kshirsagar)*

Volume 20: Handbook of Statistical Distributions, *Jagdish K. Patel, C. H. Kapadia, and D. B. Owen*

Volume 21: Case Studies in Sample Design, *A. C. Rosander*

Volume 22: Pocket Book of Statistical Tables, *compiled by R. E. Odeh, D. B. Owen, Z. W. Birnbaum, and L. Fisher*

Volume 23: The Information in Contingency Tables, *D. V. Gokhale and Solomon Kullback*

OTHER VOLUMES IN PREPARATION

The Information in Contingency Tables

D.V. GOKHALE

Department of Statistics
University of California at Riverside
Riverside, California

SOLOMON KULLBACK

Department of Statistics
The George Washington University
Washington, D.C.

MARCEL DEKKER, INC. New York and Basel

Library of Congress Cataloging in Publication Data

Gokhale, Dattaprabhakar V. [Date]
Information in contingency tables.

(Statistics, textbooks, and monographs ; v. 23)
Bibliography: p.
Includes index.
1. Contingency tables. I. Kullback, Solomon, joint author. II. Title.
QA277.G64 519.5'3 78-2761
ISBN 0-8247-6698-9

MARCEL DEKKER, INC.
270 Madison Avenue, New York, New York 10016

Current printing (last digit):
10 9 8 7 6 5 4 3

PRINTED IN THE UNITED STATES OF AMERICA

TO MY PARENTS

D. V. Gokhale

TO LOLA

S. Kullback

PREFACE

The aim of this book is to present methods of analyzing contingency tables and other types of categorical or count data not necessarily arrayed as a contingency table, by applying the principle of minimum discrimination information (MDI). Emphasis is on the methodology. The book is designed not only for statisticians and prospective statisticians, but also for professionals from any discipline which involves the analyses of categorical data presented in a cross-tabulated form such as a multi-dimensional contingency table, or count data not necessarily arrayed as a contingency table. Some knowledge of probability and statistics, including an introduction to regression and matrix notation, is desirable.

The MDI approach naturally leads to loglinear models and enables one to find estimates of cell-entries under various hypotheses or models and to test these hypotheses or models. It provides indication of outlier cells and estimates of parameters and their asymptotic covariance matrix. The structure of a contingency table can be studied in detail in terms of the various interrelationships among the classificatory variables. We believe that our approach has three important advantages. First, the approach, being based on a general principle, provides a <u>unified</u> treatment of contingency tables of any order and dimension. Step-by-step generalizations to higher order dimensional contingency tables are not necessary, as has been the case with ad hoc procedures. The MDI approach can treat univariate and multivariate logit analysis and quantal response analysis as particular cases. Second, the MDI approach is practical in that convergent iterative computer algorithms are in use and available for the prescribed analyses. This analysis technique has been successfully applied in many diverse areas. Third, the approach is reasonable in the sense that MDI estimates are, in general, best asymptotically normal. Further, for a large subclass of analyses they are maximum likelihood estimates.

In the past two or three years several books on the analysis of categorical data have been published. The present book differs from these mainly on three salient points. We have tried to keep the book as free as possible from involved theoretical proofs, the associated jungle of notational symbolism, and high powered mathematical theorems. For those readers who are interested in proofs, adequate references to literature are cited in the text. Some proofs which are felt necessary and basic are included or given in the Appendix to preserve continuity and readability. The second important feature is that all of the concepts are utilized in the numerous illustrative examples. We expect that this should enhance the reader's understanding of the subject, as one may assimilate and review all the generalities in various applications. The book is thus intended to be "user-oriented" rather than a mathematical treatise. The reader is urged to study the illustrative examples in Chapters 4, 7, and 8 with care. Third, we have not restricted our analyses of the loglinear models only to cases using iterative proportional fitting.

Those interested in obtaining copies of the computer programs, written using PL/1, may write to Secretary, Department of Statistics, The George Washington University, Washington, D.C. 20052. There will be a minimal charge to cover the cost of magnetic tape reel, manual, postage, and handling.

D. V. Gokhale

Solomon Kullback

ACKNOWLEDGMENTS

This book has been made possible by partial support under a number of grants, and the interest and encouragement of students, colleagues, and friends. It is the product of the interaction among many people, including students, colleagues, collaborators, interested statisticians, referees, and editors. Various drafts were used as the text for classes at The George Washington University, University of California, Riverside, and Stanford University. We are especially grateful to the students in these classes for their pertinent questions, comments, and suggestions that improved the exposition.

Professor C. T. Ireland and Dr. H. H. Ku were most helpful from the very beginnings of the development of the current techniques; their questions, insight, comments, and cooperation helped get it moving. The support of The George Washington University in providing an academic environment in which the teaching of many of the results presented herein, as well as the research in developing, expanding, and applying them, was made possible, stimulated, and encouraged, must be, and is gratefully acknowledged. In particular, the support and collaboration of Professor Harold F. Bright, Professor Jerome Cornfield, Professor Henry Solomon, Professor Herbert Solomon, Dr. Marian Fisher, and Dr. John C. Keegel have contributed to any merits this work possesses, any demerits are our responsibility.

The many examples were analyzed on the basis of computations using the facilities of the Computer Centers of The National Bureau of Standards, The George Washington University, and The University of California, Riverside. The collaboration of the authors was facilitated in part by Professor Ralph A. Bradley, Department of Statistics, Florida State University, in February 1975, and by Professor Ingram Olkin, Department of Statistics, Stanford University, in July 1976, by making available certain financial aid and secretarial facilities.

A special thanks to Mr. Langhorne P. Withers, for his enthusiastic interest in the theory, techniques, and procedures described herein. In particular his support and efforts in promoting the training of personnel in the USA Operational Test and Evaluation Agency in these procedures have contributed to their successful application.

The interest of Dr. Carlyle E. Maw of the National Institute of Education in the MDI procedure and his critical questions and analysis of problems in certain areas of educational testing have opened new areas of application for the MDI techniques.

The research program which underlies this book was in part initiated under AFOSR Grant No. 932-65, continued under Grants AFOSR-68-1513, AFOSR-72-2348, Contract No. N00014-67-A-0214-0015 under the joint sponsorship of the Army, Navy, and Air Force, ONR Contract Number N00014-67-A-00214-0016, Grant No. DAHCO 4-74-G-0164, U.S. Army Research Office, Durham, North Carolina, and Research Grant HL 15191 National Heart and Lung Institute, National Institutes of Health, Bethesda, Maryland. The preparation of the final version of the book for publication and further research investigations were in part supported by DHEW/ National Institute of Education research Grant No. NIE-G-76-0091.

To Mrs. Glenda Howell and Mrs. Lola S. Kullback for their typing, and all others who have contributed, our sincere thanks.

D. V. Gokhale

Solomon Kullback

CONTENTS

The Information in Contingency Tables

Chapter 1

INTRODUCTION

I. CATEGORICAL OR COUNT DATA

Data which result from experiments in the physical sciences and engineering are usually outcomes of controlled experiments, and the observations are expressible as the values of quantitative variables. In many other fields however, the data are seldom results of controlled experiments. In addition, the observations usually can be expressed only in qualitative or categorical terms, a yes - no, alive - dead, agree - disagree, class A - class B - class C, etc. type of response resulting in frequency or count data.

For example, an individual may be classified by sex, by race, by profession, by smoking habit, by age, by incidence of coronary heart disease. If we take observations over a sample of many such individuals, the result will be a cross-classification or multi-dimensional contingency table with as many dimensions as there are classifications.

Critics of methods for contingency table analysis of categorical or count data have maintained that most of the procedures used, at least in the past, were provided mostly in the form of global chi-square tests. Nevertheless presentations of this nature may be found in the statistical literature even in

1974. However, the use of the principle of minimum discrimination information (MDI) estimation leads naturally to exponential families, which provide multiplicative or loglinear models. We shall show that:

(1) Estimates of the cell entries under various hypotheses or models can be obtained so that fitting or smoothing of the data is achieved;

(2) The adequacy or fit of the model, or the null hypothesis, can be tested;

(3) Main effect and interaction parameters can be estimated as well as their asymptotic covariance matrix;

(4) The structure of the contingency table can be studied in detail in terms of the various interrelationships among the classificatory variables;

(5) The procedures can be applied to test hypotheses about particular parameters and linear combinations of parameters that are of special interest;

(6) The procedures provide indication of outlier cells. These outliers may cause a model to give a poor overall fit but may fit nicely to the other cells excluding the outliers;

(7) Since the procedures and concepts are based on a general principle a unified treatment of multidimensional contingency tables is possible. Sequences of generalizations (step by step) to higher order dimensional contingency tables are not necessary as has been the case with other ad hoc procedures.

(8) The procedure also provides estimates based on observed or sample tables, which satisfy certain external hypotheses as to underlying probability relations in the population tables. These

estimates also preserve the inherent properties of the observed data not constrained by the hypothesis;

(9) In general, the MDI estimates are best asymptotically normal (BAN) and in the many applications of fitting models to a table based on observed sets of marginal values (or general linear combinations of observed cell values), the MDI estimates are maximum-likelihood estimates;

(10) The test statistics are MDI statistics which are asymptotically distributed as chi-square with appropriate degrees of freedom. In the particular case of fitting models to a table based on observed sets of marginal values (or general linear combinations of observed cell values), the MDI statistics are log-likelihood ratio statistics. The (total) MDI statistic under an hypothesis can be analyzed into components under subhypotheses, since the MDI statistics are additive, as are the associated degrees of freedom. The analysis is analogous to analysis of variance and regression analysis techniques. It uses a design matrix, a set of regression parameters, explanatory variables, and Analysis of Information tables.

(11) In models which give estimates for an observed table based on sets of observed marginal totals as explanatory variables, sometimes the estimates can be expressed explicitly as products of marginal values. However, this is not generally true, and estimated cell frequencies have to be computed by an iterative procedure

by the use of a computer. For the foregoing cases which are special cases of problems we call _internal_, and problems involving tests of _external_ hypotheses on underlying population probabilities, a number of convergent iterative computer algorithms are in use and available. They provide as output, design matrices, the observed cell entries and the cell estimates as well as their logarithms, parameter estimates, outlier values, MDI statistics and their corresponding significance levels, and estimates of asymptotic covariance matrices of parameter estimates, to assist in and simplify the numerical aspects of the inference. In this respect it is of interest to cite the following quotation from a book review by D. J. Finney in _Journal of the Royal Statistical Society, Series A(General)_ Vol. 136 (1973), part 3, p. 461, "No mention is made of the extent to which computers have destroyed the need to assess statistical methods in terms of arithmetical simplicity: indeed the emphasis on avoiding lengthy, but easily programmed, iterative calculations is remarkable."

It may be of interest to remark that many of the concepts, and ideas we shall discuss in this book, were presented in a very general and more theoretical formulation in Kullback (1959).

In this chapter we shall describe and illustrate frequency or count data in the form of contingency tables, develop the notation we shall need subsequently, and consider some areas of investigation pertaining to contingency tables in terms of specific examples. In later chapters there are many examples of real data whose analysis illustrates the features mentioned above.

II. CONTINGENCY TABLES

There are two ways in which statistical data are collected. In one form, actual measurements are recorded for each individual in the sample; in the other, the individuals are classified as belonging to different categories. On many occasions classifications are used to present original data on direct measurements. A well-known example is that of a frequency distribution. Data collected in the form of measurements may later be grouped and presented as a frequency distribution. An important advantage of grouping is that it results in a considerable reduction of data. On the other hand, it is not usually possible to convert grouped or classified data back into the original form.

A contingency table is a form of presentation of grouped data. In the simplest case, a group of N items may be classified into just two groups, according to, say, presence or absence of a certain characteristic. For a fixed (given) characteristic the different groups of classification are called categories. For example, a group of N individuals may be classified according to hair-color (characteristic), the categories being black, brown, blonde and other. The categories may be qualitative as above, or may be quantitative, as for example in the classification by weight in pounds consisting of five categories: 40-80, 80-120, 120-160, 160-200, 200-240. When there is only one characteristic according to which data are classified we get a _one-way table_. If there are two ways of classification, say according to Rows and Columns, the Row-classification having r categories and the Column-classification having c categories, the table is called a _two-way table_ or a _r x c table_. The latter notation gives the number of categories in each classification. Carrying this notation further, a _r x c x d table_ will have three characteristics of classification, the first having r categories, the second having c and the third having d.

III. EXAMPLES

Example 1: The following is a one-way table with one classification characteristic (Geographic Area) and four categories. It gives the distribution of students by Geographic Area.

East	North	West	South	Total
4201	4552	2840	5130	16723

Example 2: Consider the distribution of 20 balls in six cells.

Cell	1	2	3	4	5	6	Total
Occupancy	2	4	4	5	1	4	20

It may be recalled at this point that in many situations such a distribution of N balls in k cells is adequately described by the multinomial distribution. We may therefore expect that the multinomial distribution will have an important role to play in the analysis of contingency tables.

Example 3: The distribution of students by Geographic Area (as in Ex. 1.) and Sex gives rise to the following 2 x 4 contingency table.

Sex	Geographic Area				Totals
	East	North	West	South	
Male	2201	2350	1400	3100	9051
Female	2000	2202	1440	2030	7672
Totals	4201	4552	2840	5130	16723

Note that this is called a 2 x 4 table since the Row-classification (Sex) has 2 categories. If the Geographic Areas were written in rows and the Sex were to correspond to columns we would get a 4 x 2 table. We shall follow this convention throughout.

Observe that for a two-way table there are two sets of marginal totals. In the above table the totals on the right can be looked upon as a one-way table with Sex as a characteristic and two categories, Male and Female. At the bottom of the above table, we see the one-way table of Ex. 1. This shows that any two-way table is associated with two one-way tables given by the marginal totals of each characteristic.

Example 4: The data below are octane determinations on independent samples of gasoline obtained in two regions of the northeastern United States in the summer of 1953, (Brownlee, Statistical Theory and Methodology, J. Wiley, 1965, p. 306).

Region A:	84.0	83.5	84.0	85.0	83.1	83.5	81.7
	85.4	84.1	83.0	85.8	84.0	84.2	82.2
	83.6	84.9					
Region D:	80.2	82.9	84.6	84.2	82.8	83.0	82.9
	83.4	83.1	83.5	83.6	83.7	82.6	82.4
	83.4	82.7	82.9	83.7	81.5	81.9	81.7
	82.5						

The problem of interest is whether the variability in the octane numbers could be regarded as the same for the two regions. Since the number of sample-values for Regions A and D are small (16 and 22 respectively) the data can be conveniently analyzed in the given form. For the sake of illustration, suppose that we classify the octane readings into three categories; below 83.5 as Poor, between 83.5 and 84.5 as Normal and above 84.5 as Good, we will get the following 2 x 3 table:

Region	Gasoline Quality			Totals
	Poor	Normal	Good	
A	4	8	4	16
D	16	5	1	22
Totals	20	13	5	38

This illustrates how to prepare contingency tables from actual measurement data. But the example brings out another important point. The contingency table, in fact, represents two frequency distributions, one for Region A and the other for Region D laid side by side. This table is different from the ones we came across earlier in that we did not start the classification with a total of 38 values, to be classified according to Region and Quality; rather we had *a priori* a set of 16 values for Region A and 22 values for Region D. (Further the sampling for the two regions was done independently). In other words, the set of marginal totals (on the one-way table) for Region was fixed before the experiments. Later on we shall have ample opportunities to see the effect of such restrictions on the analyses. At present, it is enough to know that tables as above may be regarded as contingency tables with fixed (restricted) marginal totals.

IV. NOTATIONS AND PRELIMINARIES

We have seen that the entries in the cells of a contingency table are counts or frequencies of occurrence. We shall denote these frequencies generically by the letter x, with or without subscripts. These frequencies are a result of the classification of a fixed number of individuals according to a certain probability distribution. Hence the observed frequencies x can be looked upon as realizations of a random variable X.

The cell of a contingency table and the observed frequency or count in that cell are symbolically associated in the following manner. In the Example 1, we have a one-way table representing the distribution of 16723 students by Geographic Area. We denote the occurrence in the table by x(i) with the notation

Characteristic	Index	1	2	3	4
Geographic Area	i	East	North	West	South

Thus x(3), for example, equals 2840. The total 16723 of all x(i) for i = 1, 2, 3 and 4, will be denoted by x(.).
That is, $\Sigma_{i=1}^{4}$ x(i) = x(.) = 16723. For the two-way table of Ex. 3, we denote the frequencies in the table by x(ij) with the notation

Characteristic	Index	1	2	3	4
Sex	i	Male	Female		
Geographic Area	j	East	North	West	South

Then $x(23) = 1440$, $x(14) = 3100$ and so on. To denote marginal totals we shall use the <u>dot notation</u> as before. The <u>row marginals</u> are

$\Sigma_{j=1}^{4} x(1j) = x(1.) = 9051$, $\Sigma_{j=1}^{4} x(2j) = x(2.) = 7672$.

The <u>column marginals</u> are

$\Sigma_{i=1}^{2} x(i1) = x(.1) = 4201, \ldots, \Sigma_{i=1}^{2} x(i4) = x(.4) = 5130$.

The grand total is denoted by $x(..)$ so that $x(1.) + x(2.) = x(..) = x(.1) + x(.2) + x(.3) + x(.4) = 16723 = N$.

Now consider the following three-way table: Propagation of plum root stocks from root-cuttings

	At once		Spring		
Response (Mortality)	Long	Short	Long	Short	Totals
Alive	156	107	84	31	378
Dead	84	133	156	209	582
Totals	240	240	240	240	960

The frequencies in the cells are denoted by $x(ijk)$ with the notation

Characteristic	Index	1	2
Mortality	i	Alive	Dead
Time of Planting	j	At once	Spring
Length of cutting	k	Long	Short

The marginals are as follows:

One-way marginals: $\Sigma_j \Sigma_k x(ijk) = x(i..)$, $i = 1, 2$,

$\Sigma_i \Sigma_k x(ijk) = x(.j.)$, $j = 1, 2$,

$\Sigma_i \Sigma_j x(ijk) = x(..k)$, $k = 1, 2$,

Two-way marginals: $\Sigma_i\ x(ijk) = x(.jk)$, $j = 1, 2$, $k = 1, 2$,

$\Sigma_j\ x(ijk) = x(i.k)$, $i = 1, 2$, $k = 1, 2$,

$\Sigma_k\ x(ijk) = x(ij.)$, $i = 1, 2$, $j = 1, 2$.

Note that $\Sigma_i\ x(ij.) = x(.j.)$, $\Sigma_j\ x(ij.) = x(i..)$, $\Sigma_i\ x(i..) = x(...)$ etc.

For the above table, $x(1..) = 378$, $x(2..) = 582$ and $x(...) = 960$. It should be observed that $x(.jk) = 240$ for all the four combinations of j and k. This restriction is imposed by the method of experimentation; for each combination of the planting time and cutting length exactly 240 root stocks were used and their mortality observed. This is another case of fixed marginals, similar to the one encountered in Ex. 4. It should however be noted that the statistical analysis we shall consider does not require the <u>same number</u> of root stocks for each of the experimental conditions.

The notation for cell frequencies or counts and for marginal totals can be extended in an obvious manner to four-way, five-way and higher order tables.

Let us now recall that in a contingency table a number of individuals are classified into cells. In other words for a <u>given</u> cell, an individual is classified in the cell with a certain probability. In a four-way table, for example, each cell will be denoted by (i, j, k, ℓ) for some values of the indices i, j, k and ℓ. The probability that an individual will be classified in this cell will be denoted by $p(ijk\ell)$. Just as we defined the marginal totals for the cell frequencies $x(ijk\ell)$ we may define marginal totals for probabilities. For example,

$$p(i...) = \Sigma_j \Sigma_k \Sigma_\ell\ p(ijk\ell),$$

$$p(.j.\ell) = \Sigma_i \Sigma_k\ p(ijk\ell),$$

etc.

For a two-way table the cell probabilities will be denoted by $p(ij)$, for a three-way table by $p(ijk)$ and so on. But we would like to develop the theory of <u>all</u> contingency tables in a unified manner. For this purpose it is necessary to use a symbol, ω, say, which will generically denote cells like (ij) in a two-way table, ($ijk\ell$) in a four-way table and so on. For example, in a 2 x 3 x 5 table, the symbol $x(\omega)$ will replace $x(ijk)$, being one of the 2 x 3 x 5 = 30 cells. The symbol ω here corresponds to the triplet (ijk) and takes "values" in lexicographic order (1, 1, 1), (1, 1, 2), ..., (1, 1, 5), (1, 2, 1), (1, 2, 2), ..., (1, 2, 5), ..., (2, 3, 1), (2, 3, 2), ..., (2, 3, 5). We shall also use Ω to denote the set of cells of a contingency table as well as their number when there is no confusion possible.

We shall look upon a contingency table either as a single sample from a multinomial distribution or as a collection of two or more samples from as many multinomial distributions. We shall assume throughout that the probability of classification into a cell remains constant from individual to individual and that the individuals are sampled independently. If the total number N of individuals is sampled from the population at large, and then classified, we get a single multinomial sample with Ω cells. On the other hand, if in a contingency table some marginal totals are fixed in advance by the sampling scheme, the data may be treated as a collection of k ($k \geq 2$) independent multinomial samples, where the i-th sample has Ω_i cells, $i = 1, 2, \ldots, k$. For example the contingency table of Example 3 is a single sample with $\Omega = 8$ cells and $N = 16723$. The table of Example 4 is a set of two ($k = 2$) independent samples with three cells each ($\Omega_1 = 3$, $\Omega_2 = 3$) and $N_1 = 16$, $N_2 = 22$. The three-way table of this section is a set of four independent binomials with $\Omega_1 = \Omega_2 = \Omega_3 = \Omega_4 = 2$ and $N_1 = N_2 = N_3 = N_4 = 240$.

We shall denote estimates of the cell entries derived by the principle of MDI estimation under various hypotheses or models by $x_\alpha^*(\omega)$, where values of the subscript α will range over the hypotheses or models.

V. INVESTIGATIONS ASSOCIATED WITH CONTINGENCY TABLES

A. Interaction

Classical problems in the historical development of the analysis of contingency tables concerned themselves primarily with hypotheses and tests of independence or conditional independence, or homogeneity or conditional homogeneity over different samples. They were similar to problems in classical (quantitative) multivariate analysis dealing with independence, multiple correlation, partial correlation, canonical correlation, etc. In contingency tables, such classical problems turn out to be special cases of the techniques we shall discuss. These techniques result in analyses which are essentially regression type analyses. As such they enable one to determine the relationship of one or more dependent variables on a set of explanatory variables, as well as to assess the relative effects of changes in the explanatory variables on the dependent variables. The object of the analyses is the study of interactions between classifications. The term interaction is used here in a general sense to cover both stochastic dependence and association (see for example, Bartlett (1935), Simpson (1951), Roy and Kastenbaum (1956), Ku, Varner, and Kullback (1971)).

In the analysis of contingency tables we are usually interested in the relationship between one classification and one or more of the other classifications. Thus in Ex. 4, on comparison of octane ratings we would like to compare the variability of the values for categories given by Regions A and D. In the three-way table of the preceding section we are interested in the relationship of mortality on time of planting and length of cutting. As another example, consider a three-way $r \times c \times d$ contingency table in which the row classification represents the response of an experiment on animals, the column classification types of treatment and the depth classification sex. The following hypotheses may be of interest.

1. Response is independent of treatment irrespective of sex.
2. Response is independent of the different combinations of treatment and sex (as against the possibility that a particular treatment is more effective in terms of the response, for a particular sex).
3. Given sex, response is independent of treatment.

Of course, not all contingency tables can be interpreted in such a straightforward manner. In some instances, all three classifications can be considered as response variables; then we may be interested in the independence or association among these responses. In other cases, a classification may be controlled, experimentally or naturally, like three specified levels of fertilizer applied or sex, in which case the classification is termed a factor. For convenience, we shall group all the concepts of association, dependence, etc. under the general term of _interaction_. No interaction between treatment and sex appears to be a more acceptable phrase than independence between treatment and sex, since the term independence is usually reserved to express the relationship between random variables. We may also say that the interaction between response and treatment does not interact with sex, meaning the degree of association between response and treatment is the same for both sexes. This concept gives rise to the idea of second-order interaction. There are a number of different approaches to the mathematical formulation and interpretation of the concept of no second-order interaction. One such approach, through the concept of generalized independence is introduced in the next section. For a study of the historical development of the concept of interaction in the analysis of multidimensional contingency tables, the following series of papers among the many that could be selected, was found to be very instructive: Bartlett (1935), Lancaster (1951), Simpson (1951), Roy and Kastenbaum (1956), Darroch (1962), Lewis (1962),

Plackett (1962, 1969), Birch (1963, 1964, 1965), Goodman (1963b, 1970, 1971), Good (1963), Kastenbaum (1965), Mantel (1966), Berkson (1968, 1972), Bhapkar and Koch (1968a, 1968b), Ku and Kullback (1968), Dempster (1971), Ku, Varner and Kullback (1971). It was pointed out by Darroch (1962), "That 'interaction' in contingency tables enjoys only a few of the fortuitously simple properties of interactions in the analysis of variance." See Kullback (1973).

Let us now see how the Hypotheses 1 - 3 stated above can be formulated symbolically. In this case the cell probabilities are denoted by $p(ijk)$, $i = 1, \ldots, r$, $j = 1, \ldots, c$, $k = 1, 2$.

1. Response is independent of treatment irrespective of sex.

Since the sex of the animal is immaterial in the statement of the hypothesis, we consider marginal totals of probabilities of the form $p(ij.)$. Now, since the response is postulated to be independent of treatment we further have

$$p(ij.) = p(i..)\, p(.j.), \quad i = 1, \ldots, r, \; j = 1, \ldots c.$$

2. Response is independent of the different combinations of treatment and sex.

The probability corresponding to a particular combination of treatment and sex is given by the marginal total $p(.jk)$. The hypothesis is formulated, therefore, as

$$p(ijk) = p(i..)\, p(.jk), \quad i = 1, \ldots, r, \; j = 1, \ldots, c, \; k = 1, 2.$$

3. Given sex, response is independent of treatment. Let the conditional probability of being classified in the cell (ijk), <u>given</u> that the individual is classified in the k-th depth classification (sex), be denoted by $p(ij|k)$. Also, the marginal conditional probability of classification in the i-th category irrespective of the column classification is $p(i.k)/p(..k)$, and a similar marginal probability for the j-th category of the column classification, given k, is $p(.jk)/p(..k)$. The hypothesis then

states that

$$p(ij|k) = p(i.k)p(.jk)/p^2(..k),\ k = 1, 2,\ i = 1, \ldots, r,\ j = 1, \ldots, c.$$

But $p(ij|k) = p(ijk)/p(..k)$ so that the above relations can be restated as

$$p(ijk) = p(i.k)p(.jk)/p(..k),\ k = 1, 2,\ i = 1, \ldots, r,\ j = 1, \ldots, c.$$

Observe that $\Sigma_i\Sigma_j\ p(ij|k) = 1$, since given that an individual fell into the k-th category, it must be classified in one of the (i,j) cells corresponding to the fixed k. This imposes the restriction that

$$\Sigma_i\Sigma_j\ p(ij|k) = 1 = \Sigma_i\Sigma_j\ p(ijk)/p(..k),\ k = 1, 2$$

i.e.

$$\Sigma_i\Sigma_j\ p(ijk) = p(..k),\ k = 1, 2.$$

B. Generalized Independence

Note that the second hypothesis (of independence) led us to the formulation $p(ijk) = p(i..)\ p(.jk)$ and the third hypothesis (of conditional independence) led to $p(ijk) = p(i.k)p(.jk)/p(..k)$. The cell probabilities in each case are expressed explicitly as products of simple functions of marginal probabilities. From another point of view, we can say that the trivariate function $p(ijk)$ is expressed as a product of (simpler) univariate and bivariate functions, of the form $p(.jk)$ and $p(i..)$, for example. When the cell probabilities are thus expressible as products of functions of a smaller subset of arguments, we say that the probabilities obey _generalized independence_. By generalized independence is meant that the cell probability of a multi-dimensional contingency table may be expressed as the product of factors which are functions of various marginals. The common notions of independence, conditional independence, homogeneity, or conditional homogeneity in contingency tables are all special cases of generalized independence.

VI. GENERAL REMARKS

A. Structural Simplicity

It should be noted that in the one sample case the complete contingency table can be specified in terms of $\Omega - 1$ probabilities. Similarly, in the k-sample case the complete table can be specified in terms of $\Sigma_{i=1}^{k}(\Omega_i - 1)$ probabilities. These numbers, either $\Omega - 1$ or $\Sigma(\Omega_i - 1)$ may be large. An objective of analyzing contingency tables is to examine whether the data are consistent with a <u>simpler meaningful structure</u> obeyed by the large number of individual probabilities.

In many instances, such a structure is imposed by a parametric model. The smaller the number of parameters in the model, the simpler is the structure. In the one-sample case, since the table is completely determined by $\Omega - 1$ probabilities, a model with $\Omega - 1$ linearly independent parameters should yield any given set of probabilities of a multinomial distribution. Simplicity of structure is indicated when the observed distribution can be satisfactorily approximated by a model containing a number of parameters smaller than $\Omega - 1$.

As an example consider the following model for a 2 x 3 contingency table with classifications A and B:

$$\begin{aligned}
\ell n(p(11)/\pi(11)) &= L + \tau_1^A + \tau_1^B + \tau_{11}^{AB} \\
\ell n(p(12)/\pi(12)) &= L + \tau_1^A + \tau_2^B + \tau_{12}^{AB} \\
\ell n(p(13)/\pi(13)) &= L + \tau_1^A \\
\ell n(p(21)/\pi(21)) &= L + \tau_1^B \\
\ell n(p(22)/\pi(22)) &= L + \tau_2^B \\
\ell n(p(23)/\pi(23)) &= L
\end{aligned} \tag{1}$$

where $\pi(ij)$, $i = 1, 2$, $j = 1, 2, 3$, is an arbitrary but known probability distribution over six cells. The interpretation of the model and of the subscripts and superscripts on the τ's will be discussed later on. At the moment, it will suffice to observe that there are essentially <u>five</u> linearly independent parameters,

τ_1^A, τ_1^B, τ_2^B, τ_{11}^{AB} and τ_{12}^{AB}. The parameter L is just a normalizing factor, so that the probabilities p(ij) sum to unity. Given any particular values to the five parameters a probability distribution, p(ij), i = 1, 2, j = 1, 2, 3, is uniquely determined. Conversely, for any distribution p(ij) (such as an observed distribution) with positive entries in each cell, one can solve the equations (1) to find τ_1^A, τ_1^B, τ_2^B, τ_{11}^{AB} and τ_{12}^{AB}.

In the present example, simplicity of structure will be introduced if we assume, say, that $\tau_{11}^{AB} = \tau_{12}^{AB} = 0$. Then the model has only three linearly independent parameters τ_1^A, τ_1^B and τ_2^B. In fact under this model, if we further assume that the distribution π satisfies either

$$\pi(ij) = 1/6,\ i = 1, 2,\ j = 1, 2, 3, \text{ or}$$

$$\pi(11)/\pi(13) = \pi(21)/\pi(23),\ \pi(12)/\pi(13) = \pi(22)/\pi(23),$$

then it can be easily shown that

$$p(ij) = p(i.)p(.j)\ i = 1, 2,\ j = 1, 2, 3,$$

corresponding to mutual independence of the two classifications A and B. Of course, it may not always be possible to relate a model with a smaller number of parameters to some hypothesis of independence or conditional independence, or homogeneity or conditional homogeneity.

The representation (1) is known as a loglinear model. It is frequently used in the literature with different parameterizations on an ad hoc basis. We do not assume such a model to start with, as others have, but derive it in chapter 3 by the principle of minimum discrimination information estimation (Birch, 1963; Bishop, 1967, 1969; Goodman, 1970; Mantel, 1966).

B. Exploratory Analysis

For a general contingency table, many hypotheses like the ones stated above may be of interest. An exploratory sequence of hypotheses may be necessary to determine an acceptable model. The analysis may then envisage one or more of the following aspects:

(1) Estimation of cell entries under various

hypotheses involving fitting sets of marginals, that is, requiring that some set of marginals of the estimated table have the same values as the corresponding marginals of the observed table.

(2) Estimation of parameters of the fitted model as well as their asymptotic covariance matrix.

(3) Estimation of cell entries under a model which satisfies some <u>external</u> constraints, that is, subject to a set of linear constraints on the underlying hypothetical probabilities.

(4) Estimation of cell entries under a model which satisfies some <u>internal</u> constraints, that is, fitting an observed table by finding an estimate such that certain linear combinations of estimated cell entries have the same values as the corresponding linear combinations of the observed cell counts. Note that (1) above is a particular case.

(5) Testing adequacy or fit of the model or the null hypothesis.

(6) Testing hypotheses about particular parameters of the model and/or about their linear combinations.

In general, the MDI estimates for large samples are best asymptotically normal (BAN) and in the many applications corresponding to internal constraints the MDI estimates in particular are maximum likelihood estimates.

The test statistics are MDI statistics which are asymptotically (for large samples) distributed as chi-square with appropriate degrees of freedom. In the case of internal constraints as above the MDI statistics are log-likelihood ratio statistics. The MDI statistic under an hypothesis can be analyzed into components each under a subhypothesis. The components, also MDI statistics, are additive, as are the associated degrees of freedom. The analysis of information is analogous to analysis of variance and regression analysis techniques, in that it involves a design matrix, a set of regression parameters, and associated explanatory variables.

Sometimes, estimates of cell-frequencies and/or underlying parameters, obtained by using a method of estimation such as maximum likelihood or MDI or minimum chi-square, can be expressed in a closed form in terms of the observed frequencies. However, this is not generally true. Then these estimates are generally computed by an iterative procedure and the use of a computer becomes necessary. For problems involving internal or external constraints as above, a number of convergent iterative computer programs are in use and available for the variety of MDI analyses of interest. These computer programs provide as output, design matrices, observed cell entries and estimated cell entries as well as their logarithms, estimates of asymptotic covariance matrices of parameter estimates, outlier values and MDI statistics with their corresponding significance levels.

C. Data Analysis

Conclusions drawn from contingency tables may be only exploratory in nature. One of the difficulties can be the availability of meaningful and reliable data. The first problem one faces in the analysis of cross-classified data is the decision on the number of classifications to be included and the categories within each classification. Typical among the problems in the analysis is how to segregate the effect on the response of some of the background variables, individually or jointly, from that of the others that are of particular interest. The data analytic attitude is empirical rather than theoretical. A more empirical attitude is natural when detailed theoretical understanding is unavailable. Estimation of parameters in fitted models and an analysis of the accuracy of the estimates should be considered more as data calibrating devices which make it easier to conceive of noisy data in terms of smooth distributions and relations, rather than necessarily as attempts to discover underlying truths. With a given data set, a variety of models may be tried on, and one selected on the ground of looks and fit. See Dempster (1971), Scheuren and Oh (1975).

Consider, for example, an experiment performed to compare the effectiveness of safety release devices for refrigerators in relation to children's safety. Children between two to five years of age are induced to crawl into refrigerators equipped with six different types of release devices. If a child can open the door of the refrigerator, from inside, within a certain time period, the response is classified as a success, otherwise a failure. The background variables studied included age, sex, weight, socio-economic status of parents. The experimental variable was one of six devices. (A partial analysis of this data may be found in Kullback, Kupperman, and Ku, 1962b, p. 581.) Some balancing of the background variables was achieved.

In other instances none of the factors are subject to experimental control, and whatever available data can be collected is reported. The analysis of this type of data, though it may only be seeking preliminary information can be important in fields of health and safety. The uncontrolled experimental data are sometimes the only realistic data available when these data deal with life, death, health, and safety, and some of these factors and responses are only expressible in qualitative terms, in the present state of art.

It is expected that the number of problems calling for the techniques of the analysis of multidimensional contingency tables will increase. Experience at The George Washington University with such a growing demand confirms this. The examination and interpretation of data from social phenomena, housing, psychology, education, environmental problems, health, safety, manpower, business, experimental testing of devices, military research and development, etc., are potential source areas.

D. Preview

Following this general introduction we shall consider further aspects of the analysis of multidimensional contingency tables in greater expository detail. We then present an introduction to MDI estimation, the loglinear representation, associated design matrices, parameters, and covariance matrices, without detailed mathematical proofs. This will enable the reader then to study the many illustrative examples that follow and present various aspects of the possible analyses. Some of the mathematical details may be found in the references.

Chapter 2

HYPOTHESES OF INDEPENDENCE IN TWO- AND THREE-WAY TABLES

This chapter deals with different hypotheses of mutual and conditional independence that can be formulated in two-way and three-way contingency tables (see, for example, Kullback, 1959, chapter 8). The estimates of cell entries are obtained under these hypotheses and the marginal constraints they satisfy in each case are studied. It should be noted that both, the maximum-likelihood and the MDI approach, give the same estimates under the hypotheses considered here. For the hypothesis of mutual independence in a two-way table, no examples are given because it is treated in many elementary text books; the readers may already be familiar with a number of examples. We reexamine the 2 x 2 table in chapter 3.

I. TWO-WAY TABLES

A. 2 x 2 Tables

For two-way 2 x 2 tables a primary question of interest is whether the row and column variables (classifications) are independent. An example of such a table is shown in Table 1.

TABLE 1

Observed Values x(ij)

	j = 1	j = 2	
i = 1	x(11)	x(12)	x(1.)
i = 2	x(21)	x(22)	x(2.)
	x(.1)	x(.2)	x(..) = n

To answer this question one estimates the cell entries under the hypothesis of independence as a product of the marginals, that is, denoting the estimate by $x^*(ij)$ one uses $x^*(ij) = x(i.)x(.j)/n$. Some appropriate measure of the deviation between x(ij) and $x^*(ij)$ is then used to determine whether the differences are larger than one would reasonably expect under the hypothesis of independence.

The estimated two-way table under the hypothesis or model of independence is given in Table 2.

TABLE 2

Estimate Under Independence $x^*(ij)$

	j = 1	j = 2	
i = 1	x(1.)x(.1)/n	x(1.)x(.2)/n	x(1.)
i = 2	x(2.)x(.1)/n	x(2.)x(.2)/n	x(2.)
	x(.1)	x(.2)	n

Note that the estimated table has the same marginals as the observed table x(ij).

B. Cross-product Ratio

A common statistical measure of the association or interaction between the variables (classifications) of a two-way 2 x 2 contingency table is the cross-product ratio, or its logarithm. The cross-product ratio is defined by $x(11)x(22)/x(12)x(21)$, though we shall be more concerned with its logarithm $\ln x(11)x(22)/x(12)x(21)$. We shall use natural logarithms, that is, logarithms to the base e, rather than common logarithms to the base 10, because of the nature of the underlying mathematical statistical theory. Note that with the estimate for independence, or no association, the logarithm of the cross-product ratio is zero

$$\ln \frac{x^*(11)x^*(22)}{x^*(12)x^*(21)} = \ln \frac{(x(1.)x(.1)/n)(x(2.)x(.2)/n)}{(x(1.)x(.2)/n)(x(2.)x(.1)/n)} = \ln 1 = 0.$$

The logarithm of the cross-product ratio is positive if the odds satisfy the inequalities $x(11)/x(21) > x(12)/x(22)$ or $x(11)/x(12) > x(21)/x(22)$, since then we get for the log-odds

$$\begin{aligned} \ln x(11)x(22)/x(12)x(21) &= \ln x(11)/x(21) - \ln x(12)/x(22) > 0 \\ &= \ln x(11)/x(12) - \ln x(21)/x(22) > 0. \end{aligned}$$

The logarithm of the cross-product ratio is negative if the odds satisfy the inequalities $x(11)/x(21) < x(12)/x(22)$ or $x(11)/x(12) < x(21)/x(22)$, since then we get for the log-odds

$$\begin{aligned} \ln x(11)x(22)/x(12)x(21) &= \ln x(11)/x(21) - \ln x(12)/x(22) < 0 \\ &= \ln x(11)/x(12) - \ln x(21)/x(22) < 0. \end{aligned}$$

The logarithm of the cross-product ratio thus varies from $-\infty$ to $+\infty$. Later we shall consider procedures for assessing the significance of the deviation of the logarithm of the cross-product ratio from zero, the value corresponding to no association or no interaction.

C. r x c Tables

Similar procedures apply to the case of a two-way r x c contingency table, that is, one with r rows and c columns.

TABLE 3

Two-way r x c Contingency Table

	1	2	...	c	.
1	x(11)	x(12)	...	x(1c)	x(1.)
2	x(21)	x(22)	...	x(2c)	x(2.)
⋮	...	...	...	...	...
r	x(r1)	x(r2)	...	x(rc)	x(r.)
	x(.1)	x(.2)	...	x(.c)	n

Under a hypothesis or model of independence of row and column classifications the estimate $x^*(ij) = x(i.)x(.j)/n$. Even if the row categories, say, are not randomly observed but selected with respect to some characteristic, the mathematical procedures are still the same for determining whether the column categories are homogeneous over the row categories. In the latter case we may consider the two-way table as a set of one-way tables. Terms which cover both the case of independence and homogeneity are association or interaction, that is, we question whether there is association or interaction among the variables (classifications).

The estimated two-way r x c contingency table under the hypothesis or model of independence is given in Table 4.

TABLE 4

Estimate Under Independence: $x^*(ij)$

	1	2		c	
1	x(1.)x(.1)/n	x(1.)x(.2)/n	...	x(1.)x(.c)/n	x(1.)
2	x(2.)x(.1)/n	x(2.)x(.2)/n	...	x(2.)x(.c)/n	x(2.)
⋮	...	...	...	...	...
r	x(r.)x(.1)/n	x(r.)x(.2)/n	...	x(r.)x(.c)/n	x(r.)
	x(.1)	x(.2)	...	x(.c)	n

Note that the estimated Table 4 has the same marginals as the observed Table 3.

The concept of log-odds can be extended to the case when the dependent variable, say the column variable, has more than two categories ($c > 2$) in the following manner: For each fixed category i of the independent variable, there are $c - 1$ log-odds given by $\ln x(i1)/x(ic)$, $\ln x(i2)/x(ic)$, ..., $\ln x(i, c - 1)/x(ic)$. The choice of $x(ic)$ in the denominator is quite arbitrary; we could have chosen any $x(ik)$ for comparison with the remaining $x(ij)$, $j \neq k$. A further comparison of log-odds of the form $\ln(x(ij)/x(im))$ can be done by considering $\ln(x(ij)/x(ic))$ and $\ln(x(im)/x(ic))$.

II. THREE-WAY TABLES

A three-way contingency table arises when each observation has three classifications with different possible numbers of categories for each classification. The simplest three-way contingency table is 2 x 2 x 2, that is, with two categories for each classification. In the general notation we have Table 5.

TABLE 5

2 x 2 x 2 Contingency Table

	i = 1		i = 2		
	j = 1	j = 2	j = 1	j = 2	
k = 1	x(111)	x(121)	x(211)	x(221)	x(..1)
k = 2	x(112)	x(122)	x(212)	x(222)	x(..2)
	x(11.)	x(12.)	x(21.)	x(22.)	n

The two-way marginals are

$$x(11.) = x(111) + x(112),$$
$$x(12.) = x(121) + x(122),$$
$$x(21.) = x(211) + x(212),$$
$$x(22.) = x(221) + x(222),$$
$$x(1.1) = x(111) + x(121),$$

$$x(1.2) = x(112) + x(122),$$
$$x(2.1) = x(211) + x(221),$$
$$x(2.2) = x(212) + x(222),$$
$$x(.11) = x(111) + x(211),$$
$$x(.12) = x(112) + x(212),$$
$$x(.21) = x(121) + x(221),$$
$$x(.22) = x(122) + x(222).$$

The one-way marginals are

$$x(1..) = x(111) + x(112) + x(121) + x(122) = x(11.) + x(12.),$$
$$x(2..) = x(211) + x(212) + x(221) + x(222) = x(21.) + x(22.),$$
$$x(.1.) = x(111) + x(112) + x(211) + x(212) = x(11.) + x(21.),$$
$$x(.2.) = x(121) + x(122) + x(221) + x(222) = x(12.) + x(22.),$$
$$x(..1) = x(111) + x(121) + x(211) + x(221) = x(1.1) + x(2.1),$$
$$x(..2) = x(112) + x(122) + x(212) + x(222) = x(1.2) + x(2.2).$$

The entries $x(ijk)$ in Table 5 may also be considered as three-way marginals. See Table 6.

A. Estimates Under Independence

With more variables there are more possible questions of interest. One may be interested in whether any pair of the variables are independent or show no interaction or association. One may be interested in conditional independence, that is, whether a pair of variables are independent given the third variable. One may be interested in whether the three variables are mutually independent or whether one of the variables is independent of the pair of the other variables. The related hypotheses were considered in chapter 1 section V. These questions of independence, no interaction or association are all answered by considering estimates which are explicitly represented in terms of products of simple functions of various marginals. We list some of these estimates. (See Kullback, 1959, Chap. 8.)

Mutual independence of i, j, and k,

$$x_1^*(ijk) = x(i..)x(.j.)x(..k)/n^2,$$

Independence of i and (jk) jointly,

$$x_a^*(ijk) = x(i..)x(.jk)/n,$$

Conditional independence of i and j given k,

$$x_b^*(ijk) = x(i.k)x(.jk)/x(..k).$$

As might be expected, these estimates also apply in the general three-way r x s x t contingency table.

We note that the estimate under mutual independence of i, j, and k has the same one-way marginals as the observed table x(ijk):

$$x_1^*(111) = x(1..)x(.1.)x(..1)/n^2,$$

$$x_1^*(112) = x(1..)x(.1.)x(..2)/n^2,$$

$$x_1^*(121) = x(1..)x(.2.)x(..1)/n^2,$$

$$x_1^*(122) = x(1..)x(.2.)x(..2)/n^2,$$

$$x_1^*(211) = x(2..)x(.1.)x(..1)/n^2,$$

$$x_1^*(212) = x(2..)x(.1.)x(..2)/n^2,$$

$$x_1^*(221) = x(2..)x(.2.)x(..1)/n^2,$$

$$x_1^*(222) = x(2..)x(.2.)x(..2)/n^2,$$

$$x_1^*(1..) = x_1^*(111) + x_1^*(112) + x_1^*(121) + x_1^*(122)$$

$$= x(1..)x(.1.)/n + x(1..)x(.2.)/n$$

$$= x(1..),$$

$$x_1^*(2..) = x_1^*(211) + x_1^*(212) + x_1^*(221) + x_1^*(222)$$

$$= x(2..)x(.1.)/n + x(2..)x(.2.)/n$$

$$= x(2..),$$

$$x_1^*(.1.) = x_1^*(111) + x_1^*(112) + x_1^*(211) + x_1^*(212)$$

$$= x(1..)x(.1.)/n + x(2..)x(.1.)/n$$

$$= x(.1.),$$

$$x_1^*(.2.) = x_1^*(121) + x_1^*(122) + x_1^*(221) + x_1^*(222)$$

$$= x(.2.),$$

$$x_1^*(..1) = x_1^*(111) + x_1^*(121) + x_1^*(211) + x_1^*(221)$$
$$= x(..1),$$
$$x_1^*(..2) = x_1^*(112) + x_1^*(122) + x_1^*(212) + x_1^*(222)$$
$$= x(..2).$$

However, the two-way marginals of the estimate under mutual independence of i, j, and k differ from the two-way marginals of the observed table x(ijk). Thus, for example,

$$x_1^*(11.) = x_1^*(111) + x_1^*(112)$$
$$= x(1..)x(.1.)x(..1)/n^2 + x(1..)x(.1.)x(..2)/n^2$$
$$= x(1..)x(.1.)/n,$$

and the latter value is not necessarily equal to x(11.).

The estimate under the hypothesis or model of independence of i and (jk) jointly has the same one-way marginals and the same two-way jk-marginal as the observed table x(ijk),

$$x_a^*(111) = x(1..)x(.11)/n,$$
$$x_a^*(112) = x(1..)x(.12)/n,$$
$$x_a^*(121) = x(1..)x(.21)/n,$$
$$x_a^*(122) = x(1..)x(.22)/n,$$
$$x_a^*(211) = x(2..)x(.11)/n,$$
$$x_a^*(212) = x(2..)x(.12)/n,$$
$$x_a^*(221) = x(2..)x(.21)/n,$$
$$x_a^*(222) = x(2..)x(.22)/n,$$
$$x_a^*(1..) = x_a^*(111) + x_a^*(112) + x_a^*(121) + x_a^*(122)$$
$$= x(1..)x(.11)/n + x(1..)x(.12)/n$$
$$+ x(1..)x(.21)/n + x(1..)x(.22)/n$$
$$= x(1..)(x(.11) + x(.12) + x(.21) + x(.22))/n$$
$$= x(1..).$$

Similar results follow for the other one-way marginals.

$$x_a^*(.11) = x_a^*(111) + x_a^*(211)$$
$$= x(1..)x(.11)/n + x(2..)x(.11)/n$$
$$= x(.11),$$

$$x_a^*(.12) = x_a^*(112) + x_a^*(212)$$
$$= x(1..)x(.12)/n + x(2..)x(.12)/n$$
$$= x(.12),$$

$$x_a^*(.21) = x_a^*(121) + x_a^*(221)$$
$$= x(1..)x(.21)/n + x(2..)x(.21)/n$$
$$= x(.21),$$

$$x_a^*(.22) = x_a^*(122) + x_a^*(222)$$
$$= x(1..)x(.22)/n + x(2..)x(.22)/n$$
$$= x(.22).$$

However, for the other two-way marginals, for example,

$$x_a^*(11.) = x_a^*(111) + x_a^*(112)$$
$$= x(1..)x(.11)/n + x(1..)x(.12)/n$$
$$= x(1..)(x(.11) + x(.12))/n$$
$$= x(1..)x(.1.)/n,$$

and the latter value is not necessarily equal to $x(11.)$.

$$x_a^*(1.1) = x_a^*(111) + x_a^*(121)$$
$$= x(1..)x(.11)/n + x(1..)x(.21)/n$$
$$= x(1..)(x(.11) + x(.21))/n$$
$$= x(1..)x(..1)/n,$$

and the latter value is not necessarily equal to $x(1.1)$.

The estimate under the hypothesis or model of conditional independence of i and j given k has the same one-way marginals and the same two-way ik- and jk-marginals as the observed table $x(ijk)$,

$$x_b^*(111) = x(1.1)x(.11)/x(..1),$$

$$x_b^*(112) = x(1.2)x(.12)/x(..2),$$

$$x_b^*(121) = x(1.1)x(.21)/x(..1),$$

$$x_b^*(122) = x(1.2)x(.22)/x(..2),$$

$$x_b^*(211) = x(2.1)x(.11)/x(..1),$$

$$x_b^*(212) = x(2.2)x(.12)/x(..2),$$

$$x_b^*(221) = x(2.1)x(.21)/x(..1),$$

$$x_b^*(222) = x(2.2)x(.22)/x(..2),$$

$$\begin{aligned} x_b^*(1..) &= x_b^*(111) + x_b^*(112) + x_b^*(121) + x_b^*(122) \\ &= x(1.1)x(.11)/x(..1) + x(1.2)x(.12)/x(..2) \\ &\quad + x(1.1)x(.21)/x(..1) + x(1.2)x(.22)/x(..2) \\ &= x(1.1) + x(1.2) = x(1..). \end{aligned}$$

Similar results follow for the other one-way marginals.

$$\begin{aligned} x_b^*(1.1) &= x_b^*(111) + x_b^*(121) \\ &= x(1.1)x(.11)/x(..1) + x(1.1)x(.21)/x(..1) \\ &= x(1.1), \end{aligned}$$

$$\begin{aligned} x_b^*(1.2) &= x_b^*(112) + x_b^*(122) \\ &= x(1.2)x(.12)/x(..2) + x(1.2)x(.22)/x(..2) \\ &= x(1.2), \end{aligned}$$

and in a similar manner we have

$$x_b^*(2.1) = x(2.1),\ x_b^*(2.2) = x(2.2),$$

$$\begin{aligned} x_b^*(.11) &= x_b^*(111) + x_b^*(211) \\ &= x(1.1)x(.11)/x(..1) + x(2.1)x(.11)/x(..1) \\ &= x(.11), \end{aligned}$$

$$\begin{aligned} x_b^*(.12) &= x_b^*(112) + x_b^*(212) \\ &= x(1.2)x(.12)/x(..2) + x(2.2)x(.12)/x(..2) \\ &= x(.12), \end{aligned}$$

and in a similar manner we have

$$x_b^*(.21) = x(.21),\ x_b^*(.22) = x(.22).$$

However, for the other two-way marginals

$$x_b^*(11.) = x_b^*(111) + x_b^*(112)$$

$$= x(1.1)x(.11)/x(..1) + x(1.2)x(.12)/x(..2),$$

and the latter value is not necessarily equal to x(11.).

We remark that one of the constraints in the determination of the estimates was that they have certain marginals the same as the observed table.

To illustrate the preceding discussion we give in Table 6 observations on 46 subjects showing their reactions to three drugs where the occurrence in the table is denoted by x(ijk) with the notation

Characteristic	Index	1	2
Drug A	i	Favorable	Unfavorable
Drug B	j	Favorable	Unfavorable
Drug C	k	Favorable	Unfavorable

In Table 7 are listed the original observations in lexicographic order as well as the estimates $x_1^*(ijk)$, $x_a^*(ijk)$, $x_b^*(ijk)$ corresponding to mutual independence of the reactions to the drugs, independence of the reaction to drug A and the joint reaction to drugs B and C, conditional independence of the reaction to drugs A and B given the reaction to drug C. These data have been used for other analyses by Bishop, Fienberg, and Holland (1975), Grizzle, Starmer, and Koch (1969), Koch and Reinfurt (1971).

B. No Second-order Interaction

For the three-way 2 x 2 x 2 contingency table in addition to the classic type of independence, interaction or association, there arises an additional one, important historically and practically. This is known as no three-factor or no second-order interaction. A convenient formulation of no three-factor or no second-order interaction is that the logarithm of the association measured by the cross-product ratio for any two of the variables

TABLE 6

Reactions to Drugs

	i = 1		i = 2		
	j = 1	j = 2	j = 1	j = 2	
k = 1	6	2	2	6	16
k = 2	16	4	4	6	30
	22	6	6	12	46

x(1..) = 28	x(.1.) = 28	x(..1) = 16
x(2..) = 18	x(.2.) = 18	x(..2) = 30
x(11.) = 22	x(1.1) = 8	x(.11) = 8
x(12.) = 6	x(1.2) = 20	x(.12) = 20
x(21.) = 6	x(2.1) = 8	x(.21) = 8
x(22.) = 12	x(2.2) = 10	x(.22) = 10

TABLE 7

MDI Estimates Under Independence

i	j	k	x(ijk)	$x_1^*(ijk)$	$x_a^*(ijk)$	$x_b^*(ijk)$
1	1	1	6	5.928	4.870	4.000
1	1	2	16	11.115	12.174	13.333
1	2	1	2	3.811	4.870	4.000
1	2	2	4	7.146	6.087	6.667
2	1	1	2	3.811	3.130	4.000
2	1	2	4	7.146	7.826	6.667
2	2	1	6	2.450	3.130	4.000
2	2	2	6	4.594	3.913	3.333

is the same for all the values of the third variable, that is, there is no second-order interaction if

$$
\begin{aligned}
&\ln \frac{x(111)x(221)}{x(121)x(211)} = \ln \frac{x(112)x(222)}{x(122)x(212)}, \text{ i and j for k,} \\
(1) \qquad &\ln \frac{x(111)x(212)}{x(112)x(211)} = \ln \frac{x(121)x(222)}{x(122)x(221)}, \text{ i and k for j,} \\
&\ln \frac{x(111)x(122)}{x(112)x(121)} = \ln \frac{x(211)x(222)}{x(212)x(221)}, \text{ j and k for i.}
\end{aligned}
$$

One is concerned with the possible hypothesis or model of no second-order interaction when none of the other types of independence are found. <u>However, in this case, the corresponding estimate cannot be expressed explicitly in terms of observed marginals although the estimate is constrained to have the same two-way marginals as the observed table</u>. Straighforward iterative procedures exist to determine the estimate under the hypothesis or model of no second-order interaction. For the general three-way r x s x t contingency table there are of course many more relations among the logarithms of the cross-product ratios like (1) which must be satisfied, but the iterative procedures to determine the estimate extend to the general case with no difficulty.

C. Remarks

We may be concerned with a set of two-way tables for which it is of interest to determine whether they are homogeneous with respect to a third factor, say space or time. Such problems may also be treated as three-way contingency tables using the space or time factor as the third classification (Kullback, 1959).

For four-way and higher order contingency tables the problem of presentation of the data increases, as do the variety and number of questions about relationships of possible interest and varieties of interaction. The basic ideas, concepts, notation and terminology we have discussed for the two- and three-way contingency tables extend to the more general cases as we consider the methodology.

Chapter 3

ANALYSIS BY FITTING MARGINALS

This chapter and Chapter 5 describe the principle of minimum discrimination information (MDI) estimation as applied to the analysis of contingency tables. We first state the problem in general terms and show how it leads to the _loglinear representation_ of the probabilities. We then turn to a wide subclass of problems of smoothing or fitting a model to an observed contingency table. Here the constraints specify that the estimated contingency table have some set of marginals, or more generally, linear functions of estimated cell entries, equal to the values of the same linear functions of observed cell entries. This problem is called an _internal constraints problem_ (ICP). In Chapter 5 we shall consider a class of problems we call _external constraints problems_ (ECP).

I. LOGLINEAR REPRESENTATION

Suppose there are two probability distributions or contingency tables (we shall use these terms interchangeably) defined over the set of cells or space Ω, say $p(\omega)$, $\pi(\omega)$, $\sum_{\Omega} p(\omega) = 1$, $\sum_{\Omega} \pi(\omega) = 1$. We recall the discussion at the end of section IV of Chapter 1. The discrimination information is defined by

$$I(p:\pi) = \sum_{\Omega} p(\omega)\, \ln\, (p(\omega)/\pi(\omega)). \tag{1}$$

We note for the present that the distribution $\pi(\omega)$ in (1) is arbitrary. For the various applications we shall consider, a suitable choice of the distribution $\pi(\omega)$ will be made according to the problem of interest. It may be either a specified, an estimated, or an observed distribution. The distribution $p(\omega)$ ranges over, or is a member of, a family $\mathcal{P}$ of distributions of interest satisfying certain constraints.

Of the various properties of $I(p:\pi)$ we mention in particular the fact that $I(p:\pi) > 0$ and $= 0$ if, and only if, $p(\omega) = \pi(\omega)$ (Kullback, 1959), and is a measure of the deviation between the distributions.

In accordance with the principle of MDI estimation we want the value of $p(\omega)$ which minimizes (1) over the family $\mathcal{P}$ of distributions which satisfy the linearly independent constraints (using matrix notation)

$$\underset{\sim}{C}\underset{\sim}{p} = \underset{\sim}{\theta} \tag{2}$$

where the <u>design matrix</u> $\underset{\sim}{C}$ is $(r + 1) \times \Omega$, the probability matrix $\underset{\sim}{p}$ is $\Omega \times 1$, the matrix of constraining values $\underset{\sim}{\theta}$ is $(r + 1) \times 1$, and the rank of $\underset{\sim}{C}$ is $r + 1 \leq \Omega$, that is, the rows of $\underset{\sim}{C}$ are linearly independent. Following Kullback (1959, page 387) we consider a vector as a matrix consisting of a single row or column. If we denote the elements of the matrix $\underset{\sim}{C}$ by $c_i(\omega)$, $i = 0, \ldots, r$, $\omega = 1, \ldots, \Omega$, (a variation from the usual double subscript notation), then (2) is

$$\sum_{\Omega} c_i(\omega)p(\omega) = \theta_i,\ i = 0, \ldots, r. \tag{3}$$

To express the natural constraint $\sum_{\Omega} p(\omega) = 1$, we shall take $c_0(\omega) = 1$, for all ω, and $\theta_0 = 1$. By differentiation of (1) with respect to $p(\omega)$ subject to (3), using Lagrange undetermined multipliers, the minimizing distribution is found to have the representation

(4) $p^*(\omega) = \exp(\tau_0 + \tau_1 c_1(\omega) + \tau_2 c_2(\omega) + \ldots + \tau_r c_r(\omega))\,\pi(\omega)$

or

(5) $$\ln \frac{p^*(\omega)}{\pi(\omega)} = \tau_0 + \tau_1 c_1(\omega) + \tau_2 c_2(\omega) + \ldots + \tau_r c_r(\omega),$$
$$\omega = 1, \ldots, \Omega,$$

where the τ's are to be determined so that $\underset{\sim}{C}\underset{\sim}{p}^* = \underset{\sim}{\theta}$. It has been shown that $I(p:\pi)$ is a convex function of the p's for given π's, hence the procedure yields a unique minimum. We say that $p^*(\omega)$ generates an exponential family of distributions, the family of exponential type determined by $\pi(\omega)$, as the τ parameters range over their values (Kullback, 1959). The exponential distribution $p^*(\omega)$ has a dual system of parameters, the τ's which are the exponential or natural parameters, and the θ's which are the moment parameters. We examine their interrelationships subsequently (Dempster 1969, 1971). The <u>loglinear representation</u> of cell estimates is given by (5).

We remark that $\pi(\omega)$ may also have an exponential family representation and it may be noted from (4) that those natural parameters in the exponential family representation of $\pi(\omega)$, not consistent with the moment constraints which determine $p^*(\omega)$ are carried over or inherited by $p^*(\omega)$. This property is useful in determining the selection of the $\pi(\omega)$ distribution to be used in (1). A proper selection may improve the speed of convergence in computer algorithms in ICP's or ECP's. It may also be used to retain some specified natural parameter values in the loglinear representation of $p^*(\omega)$ (see Ex. 6, Chapter 4).

The MDI estimate is $p^*(\omega)$, so that

$$I(p^*:\pi) = \Sigma\, p^*(\omega) \ln (p^*(\omega)/\pi(\omega)) = \min I(p:\pi),\ p,\ p^* \in \mathcal{P}$$

Unless otherwise stated, the summation is over Ω which will be omitted.

It may be shown that if $p(\omega)$ is any member of the family $\mathcal{P}$ of distributions, then

(6) $$I(p:\pi) = I(p^*:\pi) + I(p:p^*).$$

The pythagorean type property (6) plays an important role in the analysis of information tables (Cencov, 1968; Kullback, 1959, Chapter 8). We leave it as an exercise for the reader to prove (6).

II. FITTING MARGINALS, INTERNAL CONSTRAINTS PROBLEM

The design matrix $\underset{\sim}{C}$ and vector $\underset{\sim}{\theta}$ in (2), though assumed to be known, can take any arbitrary values as long as the system of equations is consistent. We shall now specialize the matrix $\underset{\sim}{C}$ and the vector $\underset{\sim}{\theta}$ to problems of smoothing or fitting the contingency table, when the moment constraints for the estimate are purely internal, as explained below. In fact, to distinguish this case from the general case we replace the matrix $\underset{\sim}{C}$ by a matrix $\underset{\sim}{T}'$.

The analysis can be motivated and formulated as follows. An experiment has been designed and observations made resulting in a multidimensional contingency table with the desired classifications and categories. All the information the analyst hopes to obtain from the experiment is contained in the contingency table. The aim of the analysis is to get a good fit to the observed table using a minimal or parsimonious number of natural parameters in the loglinear model (5) depending only on some of the observed marginals, and/or some general linear combinations of observed cell entries. This shows how much of the total information is contained in a summary consisting only of sets of marginals, and/or some linear combinations of observed cell entries with respect to model (5). The observed distribution can then be said to be explainable in terms of a smaller number of linear functions of cell frequencies. Such fitting or smoothing of data problems we have called internal constraints problems (ICP).

It is interesting to note that the observed distribution and the uniform distribution are two extreme ends of a spectrum, so to speak. The observed distribution, the full or complete model, requires a maximum number of natural parameters for its description, whereas the uniform distribution requires the least. Since a desirable property of the fitted distribution is that it should contain a parsimonious number or as few natural parameters as possible, it is appropriate to start the sequential fitting procedures and the computer algorithms from the point when no natural parameters, apart from a known constant are necessary, that is, the uniform distribution.

A. MDI Statistic

To test whether an observed contingency table is consistent with the null hypothesis, or model, as represented by the MDI estimate, we compute a measure of the deviation between the observed distribution and the appropriate estimate by the MDI statistic. For notational and computational convenience, let us denote the estimated contingency table in terms of occurrences by $x^*(\omega) = np^*(\omega)$ where n is the total number of occurrences. For the smoothing or fitting class of problems, that is, the internal constraints problem (ICP) with the moment constraints implied by a set of observed marginals or more generally, linear functions of observed cell entries, (those of a generalized independence hypothesis), the MDI statistic is

$$(7) \qquad 2I(x:x^*) = 2\Sigma\, x(\omega)\, \ell n\, x(\omega)/x^*(\omega),$$

which is asymptotically distributed as chi-square with appropriate degrees of freedom and is central under the null hypothesis.

In the case of ICP the MDI estimation procedure yields the same values as MLE. The statistic in (7) is also minus twice the logarithm of the likelihood ratio statistic but this is not necessarily true for other kinds of applications of the general theory (external constraints). Berkson (1972) gives examples of this situation.

B. Design Matrix

The relationship between the concept of independence or association and interaction in contingency tables and the role the marginals play is evidenced in the historical developments in the extensive literature on the analysis of contingency tables. In particular see the references mentioned regarding the historical development of the concept of interaction in section V A of Chapter 1.

Let us denote by $\underset{\sim}{x}$ the $\Omega \times 1$ matrix of entries $x(\omega)$ of the observed contingency table arranged in lexicographic order, and denote by $\underset{\sim}{T}$ an $\Omega \times (r + 1)$ design matrix of rank $r + 1 \leq \Omega$. We recall that $\underset{\sim}{T}' = \underset{\sim}{C}$ of (2). We denote the linearly independent columns of $\underset{\sim}{T}$ by $T_i(\omega)$, $1 \leq \omega \leq \Omega$, $0 \leq i \leq r$. The condition that the MDI estimate $x^*(\omega)$ satisfy the moment constraints that some set of marginals, and/or some general linear combination of estimated cell entries, are equal to the values of the corresponding combinations of cell entries of the observed contingency table, is written in matrix notation as

$$(8) \quad \underset{\sim}{T}'\underset{\sim}{x}^* = \underset{\sim}{T}'\underset{\sim}{x}, \text{ that is, } \underset{\sim}{C}n\underset{\sim}{p}^* = \underset{\sim}{T}'\underset{\sim}{x}^* = n\underset{\sim}{\theta} = \underset{\sim}{C}\underset{\sim}{x} = \underset{\sim}{T}'\underset{\sim}{x}.$$

Those columns of $\underset{\sim}{T}$ which imply a marginal constraint are the indicator functions of the marginals, that is, the corresponding $T_i(\omega)$ will be one or zero for any cell ω, according as the cell does or does not, enter into the marginal in question. We usually take $T_0(\omega) = 1$, for all ω, to satisfy the natural or normalizing constraint $\Sigma\, x^*(\omega) = \Sigma\, x(\omega) = n$. In accordance with (4), the MDI estimate is a member of the exponential family

$$(9) \quad x^*(\omega) = \exp(\tau_0 T_0(\omega) + \tau_1 T_1(\omega) + \ldots + \tau_r T_r(\omega))n\pi(\omega).$$

If we denote the $\Omega \times 1$ matrix whose entries are $\ell n(x^*(\omega)/n\pi(\omega))$ in lexicographic order on ω by $\ell n(\underset{\sim}{x}^*/n\underset{\sim}{\pi})$, then we have from (9) the loglinear representation

$$(10) \qquad \ell n(\underset{\sim}{x}^*/n\underset{\sim}{\pi}) = \underset{\sim}{T}\,\underset{\sim}{\tau},$$

where $\underset{\sim}{\tau}$ is the $(r + 1) \times 1$ matrix of the natural parameters $\tau_0, \tau_1, \tau_2, \ldots, \tau_r$. We set the normalizing parameter $\tau_0 = L$ and $\tau_1, \ldots, \tau_r$ are main effect and interaction parameters (regression coefficients), so that (10) may be written as

$$(11) \quad \ell n \frac{x^*(\omega)}{n\pi(\omega)} = L + \tau_1 T_1(\omega) + \ldots + \tau_r T_r(\omega), \; \omega = 1, 2, \ldots, \Omega.$$

From (9) and the normalization constraint $\Sigma x^*(\omega) = n$ we have

$$L = -\ell n\ M,\ M = \Sigma \exp(\tau_1 T_1(\omega) + \ldots + \tau_r T_r(\omega))\pi(\omega).$$

Note that the expression (11) is equivalent to the version (5) with

$$\tau_0 = L,\ c_i(\omega) = T_i(\omega),\ i = 1, \ldots, r.$$

The natural parameters in (9) are to be determined so that $x^*(\omega)$ satisfies the moment constraints (8). There are convergent iterative computer algorithms in use which yield the estimate $x^*(\omega)$ in (9) satisfying (8). When the moment constraints in (8) involve only marginals the Deming-Stephan iterative proportional fitting algorithm may be used to determine the estimate $x^*(\omega)$ satisfying (8) and then the natural parameters are determined from (11). The proportional fitting iteration may be described as successively cycling through adjustments of the marginals of interest starting with the marginals of the $\pi(\omega)$ distribution until a desired accuracy of agreement between the set of observed marginals of interest and the computed marginals has been attained. (See Chapter 6, Section III.)

In such cases, $\pi(\omega)$ may be taken as <u>any</u> distribution which satisfies some of the moment constraints contained in and implied by the moment constraints (8) under examination <u>and no others</u>. This rules out the observed distribution as a choice for $\pi(\omega)$ in this case since the observed distribution satisfies additional marginal constraints not implied by the given constraints. The uniform distribution satisfies the natural constraint $\Sigma\ \pi(\omega) = 1$, and this constraint is always contained in the moment constraints for any estimate. Hence from the point of view of computer programming convenience $\pi(\omega)$ can be taken as the uniform distribution for ICP. This avoids the introduction of extraneous natural parameters not determined by the moment constraints of interest. These cases include the classical hypotheses of independence, conditional independence, homogeneity, conditional homogeneity and interaction, all of which can be considered as instances of generalized independence.

Although $n\pi(\omega)$ a constant (the uniform distribution) could be absorbed into τ_0 or L in (11), we prefer to express it explicitly because there are cases in which $n\pi(\omega)$ is not a constant (external constraints) and the expression in (9) or (11) still applies.

In terms of the representation in (9) or (11), as an exponential family, the two extreme cases are the uniform distribution for which all natural parameters (taus) except L are zero, and the observed contingency table or distribution, the complete model, for which, for a four-way r x s x t x u contingency table for example, all $\Omega - 1 = rstu - 1$ natural parameters (taus) in addition to L are needed.

C. Covariance Matrices

Since the MDI estimate $x^*(\omega)$ in (9) is a member of an exponential family, and has the desirable statistical properties of such families, the estimated asymptotic covariance matrices of the variables $T_i(\omega)$ and the associated estimated natural parameters τ_i, i = 1, 2, ..., r, are related (Kullback, 1959), pp. 43-49). Compute $\underset{\sim}{S} = \underset{\sim}{T}'\underset{\sim}{D}\underset{\sim}{T}$, where $\underset{\sim}{T}$ is the design matrix, and $\underset{\sim}{D}$ is a diagonal matrix with entries the estimates $x^*(\omega)$ in lexicographic order. Partition the matrix $\underset{\sim}{S}$ as

$$\underset{\sim}{S} = \begin{bmatrix} \underset{\sim}{S}_{11} & \underset{\sim}{S}_{12} \\ \\ \underset{\sim}{S}_{21} & \underset{\sim}{S}_{22} \end{bmatrix}$$

where $\underset{\sim}{S}_{11}$ is 1 x 1, $\underset{\sim}{S}_{12} = \underset{\sim}{S}'_{21}$ is 1 x r, $\underset{\sim}{S}_{22}$ is r x r. The estimated asymptotic covariance matrix of the $T_i(\omega)$ is $\underset{\sim}{S}_{22.1}$ and the estimated asymptotic covariance matrix of the τ_i is $\underset{\sim}{S}_{22.1}^{-1}$, where $\underset{\sim}{S}_{22.1} = \underset{\sim}{S}_{22} - \underset{\sim}{S}_{21}\underset{\sim}{S}_{11}^{-1}\underset{\sim}{S}_{12}$. These matrices are computed and furnished as part of the computer output for some of the programs in use. In the examples are illustrations of the use of the covariance matrices in testing hypotheses about the natural parameters (taus) and computing simultaneous sets of confidence intervals. This procedure also applies to the observed contingency table.

D. Chi-square Quadratics

The natural parameters (taus) are determined from the log-linear representation or regression equations (11) as sums and differences of values of $\ell n\ x^*(\omega)$ or as linear combinations thereof. A variety of statistics have been presented in the literature for the analysis of contingency tables, which are chi-square quadratics in differences of the moment parameters or chi-square quadratics in the natural parameters, or the linear combinations of logarithms of the observed or estimated values. The principle of MDI estimation and its procedures provides a unifying relationship since such statistics are shown as quadratic approximations of the MDI statistic in section VIII and in some of the examples. We remark that the corresponding approximate X^2's are not generally additive.

We mention the approximations in terms of quadratic forms in the moment parameters, or the natural parameters, as a possible bridge to help relate the familiar procedures of classical regression analysis and the procedures proposed here. This may assist in understanding and interpreting the analysis of information tables (Kullback, 1959, Chapter 10).

III. THE 2 x 2 TABLE

It may be useful at this point to reexamine the 2 x 2 table from the point of view of the preceding discussion. The algebraic details are simple in this case and exhibit the unification of the MDI estimation procedures.

A. Iterative Fitting Procedure

Suppose we have the observed 2 x 2 Table 1.

TABLE 1

x(11)	x(12)	x(1.)
x(21)	x(22)	x(2.)
x(.1)	x(.2)	n

If we obtain the MDI estimate fitting the one-way marginals, the generalized independence hypothesis is the classical hypothesis of independence of the classifications and the MDI estimate is $x^*(ij) = x(i.)x(.j)/n$. Using the iterative scaling fitting procedure, we begin with $x^{(0)}(ij) = n/4$ in each cell and adjust the $x^{(0)}(ij)$ values by the ratios of the observed row marginals to those of $x^{(0)}(ij)$, that is,

$$x^{(1)}(ij) = x^{(0)}(ij)\,\frac{x(i.)}{n/2} = x(i.)/2$$

Then we adjust $x^{(1)}(ij)$ by the ratio of observed column marginals to the marginals of $x^{(1)}(ij)$,

$$\begin{aligned} x^{(2)}(ij) &= x^{(1)}(ij)\,\frac{x(.j)}{n/2} = \frac{x(i.)}{2}\,\frac{x(.j)}{n/2} \\ &= x(i.)x(.j)/n = x^*(ij). \end{aligned}$$

Since the row and column marginals of $x^{(2)}(ij)$ are now the same as the observed marginals, no further iterative adjustment is necessary. For fitting a 2 x 2 table to externally specified marginals see Ireland and Kullback (1968b) or Fisher's 2 x 2 table in Chapter 5.

The iterative fitting procedure above started with $x^{(0)}(ij) = n/4$, the uniform distribution, as a matter of convenience. It is also the simplest distribution satisfying the independence property. The iterative fitting procedure can be started with any distribution satisfying the independence property and the same MDI estimate will be obtained. Thus starting with $x^{(0)}(ij) = np(i.)p(.j)$ we have

$$\begin{aligned} x^{(1)}(ij) &= x^{(0)}(ij)x(i.)/x^{(0)}(i.) = np(i.)p(.j)x(i.)/np(i.) \\ &= p(.j)x(i.), \\ x^{(2)}(ij) &= x^{(1)}(ij)x(.j)/x^{(1)}(.j) = p(.j)x(i.)x(.j)/np(.j) \\ &= x(i.)x(.j)/n \end{aligned}$$

as before.

B. Loglinear Representation, Complete Model

The representation of the loglinear regression for the <u>complete model</u> is given in Table 2. In this case the estimates are the same as the observed cell entries. The entries in the columns τ_1, τ_2, τ_3 are, respectively, the values of the linearly independent indicator functions $T_1(ij)$, $T_2(ij)$, $T_3(ij)$ associated with the marginals x(1.), x(.1), x(11), and the column headed L corresponds to the normalizing factor in accordance with (11). The blank spaces in the table indicate zeroes. It is seen that

$$\Sigma\Sigma\ T_1(ij)x(ij) = x(1.),\quad \Sigma\Sigma\ T_2(ij)x(ij) = x(.1),$$

TABLE 2

Complete Model

ij	L	τ_1	τ_2	τ_3
11	1	1	1	1
12	1	1		
21	1		1	
22	1			

$\Sigma\Sigma\ T_3(ij)x(ij) = x(11)$.

We note the interpretation of Table 2, in accordance with (11), as the loglinear relations

$$
\begin{aligned}
&\ell n\ (x(11)/n\pi) = L + \tau_1 + \tau_2 + \tau_3 \\
(12)\qquad &\ell n\ (x(12)/n\pi) = L + \tau_1 \\
&\ell n\ (x(21)/n\pi) = L + \tau_2 \\
&\ell n\ (x(22)/n\pi) = L.
\end{aligned}
$$

From (12) it is found that

$$
\begin{aligned}
L &= \ell n\ (x(22)/n/4), \\
\tau_1 &= \ell n\ (x(12)/x(22)), \\
\tau_2 &= \ell n\ (x(21)/x(22)), \\
\tau_3 &= \ell n\ (x(11)x(22)/x(12)x(21)),
\end{aligned}
$$

or

$$\tau_1 = \ell n\, x(12) - \ell n\, x(22),$$

(13) $$\tau_2 = \ell n\, x(21) - \ell n\, x(22),$$

$$\tau_3 = \ell n\, x(11) + \ell n\, x(22) - \ell n\, x(12) - \ell n\, x(21).$$

The natural parameters are expressed uniquely in terms of x(ij) since this is a <u>complete model</u>. The design matrix $\underset{\sim}{T}$ is the matrix of Table 2, that is,

$$\underset{\sim}{T} = \begin{bmatrix} 1 & 1 & 1 & 1 \\ 1 & 1 & 0 & 0 \\ 1 & 0 & 1 & 0 \\ 1 & 0 & 0 & 0 \end{bmatrix}$$

C. Covariance Matrix

Define the diagonal matrix $\underset{\sim}{D}$ with main diagonal the elements x(ij) in lexicographic order, that is,

$$\underset{\sim}{D} = \begin{bmatrix} x(11) & 0 & 0 & 0 \\ 0 & x(12) & 0 & 0 \\ 0 & 0 & x(21) & 0 \\ 0 & 0 & 0 & x(22) \end{bmatrix}$$

then the estimate of the asymptotic covariance matrix of the moment parameters x(1.), x(.1), x(11), for the observed contingency table is $\underset{\sim}{S}_{22.1}$ where

$$\underset{\sim}{S} = \begin{bmatrix} \underset{\sim}{S}_{11} & \underset{\sim}{S}_{12} \\ \underset{\sim}{S}_{21} & \underset{\sim}{S}_{22} \end{bmatrix} = \underset{\sim}{T}'\underset{\sim}{D}\underset{\sim}{T},$$

$$\underset{\sim}{S}_{22.1} = \underset{\sim}{S}_{22} - \underset{\sim}{S}_{21}\underset{\sim}{S}_{11}^{-1}\underset{\sim}{S}_{12},$$

and $\underset{\sim}{S}_{11}$ is 1 x 1, $\underset{\sim}{S}_{22}$ is 3 x 3, $\underset{\sim}{S}_{21}' = \underset{\sim}{S}_{12}$ is 1 x 3. It is found that

$$\underset{\sim}{S}_{22.1} = \begin{bmatrix} \dfrac{x(1.)x(2.)}{n} & x(11) - \dfrac{x(1.)x(.1)}{n} & \dfrac{x(11)x(2.)}{n} \\ x(11) - \dfrac{x(1.)x(.1)}{n} & \dfrac{x(.1)x(.2)}{n} & \dfrac{x(11)x(.2)}{n} \\ \dfrac{x(11)x(2.)}{n} & \dfrac{x(11)x(.2)}{n} & x(11) - \dfrac{x^2(11)}{n} \end{bmatrix}$$

and the inverse matrix is

$$\underset{\sim}{S}_{22.1}^{-1} =$$

$$\begin{bmatrix} \dfrac{1}{x(12)} + \dfrac{1}{x(22)} & \dfrac{1}{x(22)} & -\dfrac{1}{x(12)} - \dfrac{1}{x(22)} \\ \dfrac{1}{x(22)} & \dfrac{1}{x(21)} + \dfrac{1}{x(22)} & -\dfrac{1}{x(21)} - \dfrac{1}{x(22)} \\ -\dfrac{1}{x(12)} - \dfrac{1}{x(22)} & -\dfrac{1}{x(21)} - \dfrac{1}{x(22)} & \dfrac{1}{x(11)} + \dfrac{1}{x(12)} + \dfrac{1}{x(21)} + \dfrac{1}{x(22)} \end{bmatrix}$$

The matrix $\underset{\sim}{S}_{22.1}^{-1}$ is a large sample estimate of the covariance matrix of the taus in (13). Similar results hold in general and for estimated tables. For a derivation of the inverse matrix above using the relationship between the moment parameters and the natural parameters see Kullback (1970, ARO-D Report 71-3). We leave it as an exercise for the reader to compute the matrices above.

Note that the logarithm of the cross-product ratio, a measure of association or interaction, appears in the course of the analysis as the value of τ_3 for the observed values x(ij).

D. Loglinear Representation, Independence Model

For x*(ij), the estimate under the hypothesis of independence, the representation as in Table 2 does not involve the last column, since x*(ij) is obtained by fitting only the one-way marginals, and hence $\tau_3 = 0$.

The loglinear relations for the estimate x*(ij) are

$$
\begin{aligned}
\ell n\ (x^*(11)/n\pi) &= L + \tau_1 + \tau_2 \\
\ell n\ (x^*(12)/n\pi) &= L + \tau_1 \\
\ell n\ (x^*(21)/n\pi) &= L + \tau_2 \\
\ell n\ (x^*(22)/n\pi) &= L.
\end{aligned} \tag{14}
$$

or

$$
\begin{aligned}
\tau_1 &= \ell n\ x^*(12) - \ell n\ x^*(22), \\
\tau_2 &= \ell n\ x^*(21) - \ell n\ x^*(22), \\
0 &= \ell n\ x^*(11) - \ell n\ x^*(12) - \ell n\ x^*(21) + \ell n\ x^*(22),
\end{aligned}
$$

where the numerical values of L, τ_1, τ_2 in (14) must of course depend on x^* and differ from the values in (12). We leave it as an exercise for the reader to compute $\underset{\sim}{S}_{22.1}$ and $\underset{\sim}{S}_{22.1}^{-1}$ for $x^*(ij)$.

E. The MDI Statistic And Approximations

The MDI statistic to test the null hypothesis or model of independence is $2I(x:x^*)$ with one degree of freedom. Since $x^*(ij) = x(i.)x(.j)/n$, it may be shown that $2I(x:x^*) = 2\Sigma\Sigma\ x(ij)\ \ell n\ (x(ij)/x^*(ij)) = 2\Sigma\Sigma\ x(ij)\ \ell n\ x(ij) + 2n\ \ell n\ n - 2\Sigma\ x(i.)\ \ell n\ x(i.) - 2\Sigma\ x(.j)\ \ell n\ x(.j)$. The exact value of $2I(x:x^*)$ is easily evaluated from the expression above using tables in Kullback (1959, Chapter 8), or Kullback, Kupperman and Ku (1962a). In this case the quadratic approximation in terms of moment parameters is

$$
(15)\quad 2I(x:x^*) \approx \left(x(11) - \frac{x(1.)x(.1)}{n}\right)^2 \left(\frac{1}{x^*(11)} + \frac{1}{x^*(12)} + \frac{1}{x^*(21)} + \frac{1}{x^*(22)}\right).
$$

Remembering that $x^*(ij) = x(i.)x(.j)/n$, the right-hand side of (15) may be shown to be

$$
(16)\quad X^2 = \Sigma\ (x(ij) - x(i.)x(.j)/n)^2/x(i.)x(.j)/n,
$$

the classical Pearson X^2-test for independence with one degree of freedom. Another chi-square test which has been proposed for the null hypothesis of no association or no interaction in the 2 x 2 table, that is, $H_0:\tau_3 = 0$, is

$$(\ell n\ x(11) + \ell n\ x(22) - \ell n\ x(12) - \ell n\ x(21))^2 \times$$

$$\left(\frac{1}{x(11)} + \frac{1}{x(12)} + \frac{1}{x(21)} + \frac{1}{x(22)}\right)^{-1},$$

or $\tau_3^2/\mathrm{Var}(\tau_3)$, which may be shown to be a quadratic approximation for $2I(x:x^*)$ in terms of the natural parameter τ_3 with the covariance matrix estimated using the observed values and not the estimated values. We remark that if the observed values are used to estimate the covariance matrix then instead of the Pearson X^2-test in (16) there is derived a version of the Neyman modified chi-square

$$X_1^2 = \Sigma\ (x(ij) - x(i.)x(.j)/n)^2/x(ij).$$

IV. ANALYSIS OF INFORMATION

The analysis of information is based on the fundamental relation (6) for MDI statistics. Specifically if $np_a^*(\omega) = x_a^*(\omega)$ is the MDI estimate corresponding to a set H_a of given marginals or moment constraints, and $x_b^*(\omega)$ is the MDI estimate corresponding to a set H_b of given marginals or moment constraints, where the set H_a is explicitly or implicitly contained in the set H_b, _and the values of all the moment constraints are derived in terms of cell entries of the observed table as in (8)_, then the basic additive relations are ($\pi(\omega)$ is the uniform distribution)

$$
\begin{aligned}
(17)\qquad & 2I(x:n\pi) = 2I(x_a^*:n\pi) + 2I(x:x_a^*) \\
& 2I(x:n\pi) = 2I(x_b^*:n\pi) + 2I(x:x_b^*) \\
& 2I(x_b^*:n\pi) = 2I(x_a^*:n\pi) + 2I(x_b^*:x_a^*) \\
& 2I(x:x_a^*) = 2I(x_b^*:x_a^*) + 2I(x:x_b^*)
\end{aligned}
$$

with corresponding additive relations for the associated degrees of freedom. The last of the relations in (17) is the one with which we shall be particularly concerned. We remark that $2I(x_b^*:x_a^*)$ is the minimum value of $2I(x:x_a^*)$, for given x_a^* (perhaps estimated earlier), over all distributions which satisfy the moment constraints of the set H_b; and $2I(x:x_b^*)$ is the minimum value of $2I(x:x_a^*)$, for given $x(\omega)$, over all distributions which satisfy the moment constraints of the set H_b, that is, (recall that the information numbers are non-negative)

$$2I(x:x_a^*) \geqq 2I(x_b^*:x_a^*), \text{ with equality if and only if } x = x_b^*,$$

$$2I(x:x_a^*) \geqq 2I(x:x_b^*), \text{ with equality if and only if } x_a^* = x_b^*.$$

Measures of the form $2I(x:x_a^*)$, that is, the comparison of an observed contingency table with an estimated contingency table, are called measures of interaction or goodness-of-fit. Measures of the form $2I(x_b^*:x_a^*)$, comparing two estimated contingency tables, are called measures of effect, that is, the effect of the moment constraints in the set H_b but not in the set H_a, or the natural parameters (taus) in x_b^* but not in x_a^*. We note that $2I(x:x_a^*)$ tests a null hypothesis that the values of the natural parameters (taus) in the representation of the observed contingency table $x(\omega)$ but not in the representation of the estimated table $x_a^*(\omega)$ are zero and the number of these natural parameters (taus) is the number of degrees of freedom. Similarly $2I(x_b^*:x_a^*)$ tests a null hypothesis that the values of the set of natural parameters (taus) in the representation of the estimated table $x_b^*(\omega)$ but not in the representation of the estimated table $x_a^*(\omega)$ are zero, and the number of these natural parameters (taus) is the number of degrees of freedom.

We summarize the additive relationships of the MDI statistics and the associated degrees of freedom in the Analysis of Information Table 3.

TABLE 3

Analysis of Information

Component due to		Information	D.F.
H_a:	Interaction	$2I(x:x_a^*)$	N_a
H_b:	Effect	$2I(x_b^*:x_a^*)$	$N_a - N_b$
	Interaction	$2I(x:x_b^*)$	N_b

Since measures of the form $2I(x:x_a^*)$ may also be interpreted as measures of the variation unexplained by the estimate x_a^*, the additive relationship leads to the interpretation of the ratio

$$\frac{2I(x:x_a^*) - 2I(x:x_b^*)}{2I(x:x_a^*)} = \frac{2I(x_b^*:x_a^*)}{2I(x:x_a^*)} \tag{18}$$

as the fraction of the variation unexplained by x_a^* accounted for by the additional moment constraints defining x_b^*. The ratio (18) is thus similar to the squared correlation coefficients associated with normal distributions (Goodman, 1970).

We remark that the values of the marginals, explicit and implicit, of the estimated table $x_a^*(\omega)$, which are contained in the set of moment constraints H_a used to generate $x_a^*(\omega)$, are the same as the values of the corresponding marginals of the observed $x(\omega)$ table and all lower order implied marginals. It may be shown that $2I(x:x_a^*)$ is approximately a chi-square quadratic in the differences between the remaining moment constraints of the $x(\omega)$ table and the corresponding ones as calculated from $x_a^*(\omega)$.

Similarly, $2I(x_b^*:x_a^*)$ is also approximately a chi-square quadratic in the differences between those additional moment constraints in H_b but not in H_a and the corresponding moment constraint values as computed from the $x_a^*(\omega)$ table. Illustrations of the foregoing are to be found in the examples of Chapters 4, 7, and 8.

V. THE 2 X 2 X 2 X 2 TABLE

A. Loglinear Representation, Complete Model

A useful graphic representation of the constraints (8) and the loglinear regression (11) is given in Table 4 for a 2 x 2 x 2 x 2 contingency table. This is the analogue of the design matrix in normal regression theory. The blank spaces in Table 4 represent zero values. The ω column is the cell identification in lexicographic order. Column 1 corresponds to L which is the normalizing factor. Each of the columns 2 to 16 represents the corresponding values of linearly independent $T(\omega)$ indicator functions, columns 2 to 5 those for one-way marginals, columns 6 to 11 those for two-way marginals, columns 12 to 15 those for three-way marginals, and column 16 that for four-way marginals. The natural parameter tau associated with the $T(\omega)$ function is given at the head of the column. The superscripts are characteristic (variable) identifications, the subscripts identify the relevant categories. The complete representation with all the columns of Table 4 generates the observed values. Thus the rows represent

$$\ell n\ (x(ijk\ell)/n\pi(ijk\ell)) = L + \tau_1^i T_1^i(ijk\ell) + \ldots + \tau_{11}^{ij} T_{11}^{ij}(ijk\ell)$$

$$+ \ldots + \tau_{111}^{ijk} T_{111}^{ijk}(ijk\ell) + \ldots + \tau_{1111}^{ijk\ell} T_{1111}^{ijk\ell}(ijk\ell),$$

where $\pi(ijk\ell)$ in the 2 x 2 x 2 x 2 case is 1/16 and the numerical values of L and the natural parameters (taus) depend on the observed values $x(ijk\ell)$. The design matrix corresponding to an estimate uses only those columns associated with the marginal constraints explicit and implicit in the fitting process. This is a reflection of the fact that higher order marginals imply certain lower order marginals, for example, the two-way marginal x(ij..) fixes, by summation over i and j, the one-way marginals x(.j..), x(i...), and the total n = x(....).

TABLE 4

Graphic Representation

ω	1	2	3	4	5	6	7	8	9	10	11	12	13	14	15	16
ijkℓ	L	τ_1^i	τ_1^j	τ_1^k	τ_1^ℓ	τ_{11}^{ij}	τ_{11}^{ik}	$\tau_{11}^{i\ell}$	τ_{11}^{jk}	$\tau_{11}^{j\ell}$	$\tau_{11}^{k\ell}$	τ_{111}^{ijk}	$\tau_{111}^{ij\ell}$	$\tau_{111}^{ik\ell}$	$\tau_{111}^{jk\ell}$	$\tau_{1111}^{ijk\ell}$
1111	1	1	1	1	1	1	1	1	1	1	1	1	1	1	1	1
1112	1	1	1	1		1	1		1			1				
1121	1	1	1		1	1		1		1			1			
1122	1	1	1			1										
1211	1	1		1	1		1	1			1			1		
1212	1	1		1			1									
1221	1	1			1			1								
1222	1	1														
2111	1		1	1	1				1	1	1				1	
2112	1		1	1					1							
2121	1		1		1					1						
2122	1		1													
2211	1			1	1						1					
2212	1			1												
2221	1				1											
2222	1															

B. Loglinear Representation, Estimates

The representation for the uniform distribution corresponds to column 1 only. The MDI estimate $x_1^*(ijk\ell)$ based on fitting the one-way marginals will use only columns 1-5. The values of L and the natural parameters (taus) for this estimate will be different from those for $x(ijk\ell)$ and depend on the estimate $x_1^*(ijk\ell)$. The representation in Table 4 implies for $x_1^*(ijk\ell)$

$$\ell n\ (x_1^*(1111)/n\pi) = L + \tau_1^i + \tau_1^j + \tau_1^k + \tau_1^\ell$$

$$\ell n\ (x_1^*(1112)/n\pi) = L + \tau_1^i + \tau_1^j + \tau_1^k$$

$$\begin{matrix} \cdot & \cdot & \cdot \\ \cdot & \cdot & \cdot \\ \cdot & \cdot & \cdot \end{matrix}$$

$$\ell n\ (x_1^*(2222)/n\pi) = L.$$

It may be shown (and we leave it as an exercise for the reader) that x_1^* can be represented explicitly in terms of the marginals as $x_1^*(ijk\ell) = x(i...)x(.j..)x(..k.)x(...\ell)/n^3$, the estimate under a null hypothesis that the classifications are mutually independent. As an exercise, show that $x_1^*(i...) = x(i...)$, $x_1^*(.j..) = x(.j..)$, $x_1^*(..k.) = x(..k.)$, $x_1^*(...\ell) = x(...\ell)$. Recall Chapter 2.

The estimate $x_2^*(ijk\ell)$ based on fitting the two-way marginals will use columns 1-11 since the two-way marginals also fix the one-way marginals. The values of L and the natural parameters (taus) for this estimate will be different from those for the observed values or other estimates and depend on the values of the estimate $x_2^*(ijk\ell)$. For the estimate fitting the two-way marginals the representation in Table 4 implies

$$\ell n\,(x_2^*(1111)/n\pi) = L + \tau_1^i + \tau_1^j + \tau_1^k + \tau_1^\ell + \tau_{11}^{ij} + \tau_{11}^{ik} + \tau_{11}^{i\ell} + \tau_{11}^{jk} + \tau_{11}^{j\ell} + \tau_{11}^{k\ell}$$

$$\ell n\,(x_2^*(1112)/n\pi) = L + \tau_1^i + \tau_1^j + \tau_1^k + \tau_{11}^{ij} + \tau_{11}^{ik} + \tau_{11}^{jk}$$

$$\vdots \qquad \vdots \qquad \vdots$$

$$\ell n\,(x_2^*(2222)/n\pi) = L.$$

The estimate x_2^* is a product of <u>functions</u> of the two-way marginals and must be obtained by more than one cycle of the iterative scaling algorithm. However, $x_2^*(ij..) = x(ij..)$, ..., $x_2^*(..jk) = x(..jk)$.

The estimate $x_3^*(ijk\ell)$ based on fitting the three-way marginals will use columns 1-15 since the three-way marginals also fix the two-way and one-way marginals. This estimate is a product of <u>functions</u> of the three-way marginals and can be obtained by the use of the iterative scaling algorithm or some of the other iterative algorithms to be discussed later. However, the values of the three-way marginals of $x_3^*(ijk\ell)$ are the same as the values of the corresponding three-way marginals of $x(ijk\ell)$. We leave it as an exercise for the reader to write the loglinear representation for $x_3^*(ijk\ell)$.

Note that in the graphic representation in Table 4 we have set all natural parameters (taus) with subscript i = 2 and/or j = 2 and/or k = 2 and/or ℓ = 2, equal to zero, by convention, to insure linear independence.

C. Analysis Of Information Table

The analysis of information table corresponding to the hierarchical fitting of $x_1^*(ijk\ell)$, $x_2^*(ijk\ell)$, $x_3^*(ijk\ell)$ is shown in Table 5.

TABLE 5

Analysis of Information

Component due to	Information	D.F.
All one-way marginals	$2I(x:x_1^*)$	11
All two-way marginals	$2I(x_2^*:x_1^*)$	6
	$2I(x:x_2^*)$	5
All three-way marginals	$2I(x_3^*:x_2^*)$	4
	$2I(x:x_3^*)$	1

$2I(x:x_1^*)$ tests the null hypothesis that the eleven natural parameters (taus) of columns 6-16 are equal to zero.

$2I(x_2^*:x_1^*)$ tests the null hypothesis that the six natural parameters (taus) of columns 6-11 are equal to zero, given that the five natural parameters (taus) of columns 12-16 are zero.

$2I(x:x_2^*)$ tests the null hypothesis that the five natural parameters (taus) of columns 12-16 are zero.

$2I(x_3^*:x_2^*)$ tests the null hypothesis that the four natural parameters (taus) of columns 12-15 are zero, given that the natural parameter (tau) of column 16 is zero.

$2I(x:x_3^*)$ tests the null hypothesis that the natural parameter (tau) of column 16 is zero.

In the examples we shall see other tests on the natural parameters.

VI. THE R X S X T X U TABLE

We now describe in some detail relations among the $T(\omega)$ functions, specifically for the case of fitting a four-way $r \times s \times t \times u$ table $p(ijk\ell)$ by marginals. These functions turn out to be a basic set of simple functions and their various products. Thus, for example, the $T(\omega)$ function associated with the one-way marginal $p(2...)$ is

$$T_2^R(ijk\ell) = 1 \qquad \text{for } i = 2, \text{ any } j, k, \ell,$$
$$= 0 \qquad \text{otherwise},$$

since

$$\Sigma\, p(ijk\ell)\, T_2^R(ijk\ell) = p(2...).$$

Similarly the $T(\omega)$ function associated with the one-way marginal $p(..3.)$, for example, is

$$T_3^T(ijk\ell) = 1 \qquad \text{for } k = 3, \text{ any } i, j, \ell,$$
$$= 0 \qquad \text{otherwise},$$

since

$$\Sigma\, p(ijk\ell)\, T_3^T(ijk\ell) = p(..3.).$$

Thus for the $r \times s \times t \times u$ table we have

(19)

$(r - 1)$ linearly independent functions

$$T_\alpha^R(ijk\ell),\ \alpha = 1, \ldots, r - 1,$$

$(s - 1)$ linearly independent functions

$$T_\beta^S(ijk\ell),\ \beta = 1, \ldots, s - 1,$$

$(t - 1)$ linearly independent functions

$$T_\gamma^T(ijk\ell),\ \gamma = 1, \ldots, t - 1,$$

$(u - 1)$ linearly independent functions

$$T_\delta^U(ijk\ell),\ \delta = 1, \ldots, u - 1.$$

For example, there are (r - 1) linearly independent functions above, since

$$\sum_{\alpha=1}^{r} \Sigma \quad T_{\alpha}^{R}(ijk\ell) = rstu.$$

We have arbitrarily selected the cell in which $i = r$, $j = s$, $k = t$, $\ell = u$ as a reference cell and have excluded the functions corresponding to $\alpha = r$, $\beta = s$, $\gamma = t$, $\delta = u$ as a matter of convenience. We could have selected any cell as the reference cell.

The $T(\omega)$ function associated with the two-way marginal p(12..) say, is $T_1^R(ijk\ell)\ T_2^S(ijk\ell)$ since from the definition of $T_1^R(ijk\ell)$ and $T_2^S(ijk\ell)$ it may be seen that

$$\begin{aligned} T_1^R(ijk\ell)T_2^S(ijk\ell) &= 1 \qquad \text{for } i = 1,\ j = 2, \text{ any } k,\ \ell, \\ &= 0 \qquad \text{otherwise,} \end{aligned}$$

and

$$\Sigma\ p(ijk\ell)\ T_1^R(ijk\ell)T_2^S(ijk\ell) = p(12..).$$

Thus the $T(\omega)$ function associated with any two-way marginal is a product of two appropriate functions of the set (19).

Similarly the $T(\omega)$ function associated with any three-way marginal will be a product of three of the appropriate functions of the set (19), for example,

$$\Sigma\ p(ijk\ell)\ T_2^R(ijk\ell)T_1^T(ijk\ell)T_3^U(ijk\ell) = p(2.13).$$

Similarly the $T(\omega)$ function associated with any four-way marginal will be a product of four of the appropriate functions of the set (19), for example,

$$\Sigma\ p(ijk\ell)\ T_2^R(ijk\ell)T_1^S(ijk\ell)T_1^T(ijk\ell)T_2^U(ijk\ell) = p(2112).$$

We note that there are a total of

$$N_1 = (r - 1) + (s - 1) + (t - 1) + (u - 1)$$

$$\begin{aligned} N_2 = {} & (r - 1)(s - 1) + (r - 1)(t - 1) + (r - 1)(u - 1) \\ & + (s - 1)(t - 1) + (s - 1)(u - 1) + (t - 1)(u - 1) \end{aligned}$$

$$N_3 = (r - 1)(s - 1)(t - 1) + (r - 1)(s - 1)(u - 1) + (r - 1)(t - 1)(u - 1) + (s - 1)(t - 1)(u - 1)$$

$$N_4 = (r - 1)(s - 1)(t - 1)(u - 1)$$

respectively of the simple linearly independent functions and their products two, three, four at a time. It may be verified that

$$rstu - 1 = N = N_1 + N_2 + N_3 + N_4.$$

These values determine the degrees of freedom in the analysis of information tables. For convenience we write $T_1^R(ijk\ell)T_2^S(ijk\ell) = T_{12}^{RS}(ijk\ell)$, $T_2^R(ijk\ell)T_1^T(ijk\ell)T_3^U(ijk\ell) = T_{213}^{RTU}(ijk\ell)$, etc.

VII. FITTING STRATEGIES

A. Order of Fitting

The following points should be noted regarding the statistical significance, or otherwise, of the interaction or effect type measures $2I(x:x_a^*)$ or $2I(x_b^*:x_a^*)$ (see Table 3 above) based on chi-square values with appropriate degrees of freedom (D.F.).

(1) The MDI statistics or Information Numbers are asymptotically distributed as chi-squares. The sample size, therefore, must be large. However, contingency tables are usually based on large number of observations, so this assumption is generally valid in practice.

(2) In seeking a good fit to data, as by fitting marginals, for example, the tests are carried out sequentially. Hence they become conditional on the outcome of the previous ones. The actual values of the Information Numbers should always be considered more as measures of departures from respective hypotheses, rather than strictly as values of test statistics. A balance should be achieved between the amount of variation explained and the number of parameters associated with the marginals fitted. The Information Number of the first estimate in the fitting sequence will usually be the basis to determine the variation explained by subsequent estimates. Accordingly, the first model

is determined by the nature of the problem under consideration. If the contingency table involves classifications all of which are random variables and the problem is essentially that of determining mutual associations and interactions similar to a correlation analysis, the first model will be that for mutual independence and the estimate fits all the one-way marginals. If the contingency table involves a dependent classification and a number of explanatory classifications, and the problem is that of determining the relationship of the dependent classification on the explanatory classifications, similar to a regression analysis, the first model will be that for homogeneity of the dependent classification over the combinations of the explanatory classifications, and the estimate fits the one-way marginal of the dependent classification and the joint multiway marginal of the explanatory classifications. Incidentally, in this last case it will be found convenient because of the lexicographic ordering to use the last index for the dependent classification. The examples in Chapter 4 are not all consistent with this hint, however.

(3) The order in which the other marginals are chosen for fitting may sometimes be indicated by the objective of the analysis, fixed marginal totals, the nature of the classifications, such as time of occurrence, cause-effect phenomena etc. Otherwise, the choice of order is arbitrary and may be governed by considerations such as previous experience, comparison with an earlier analysis by different methods and so on. The value of the Information Numbers will, of course, depend on the order of fitting. For example, in a five-way table consider two possible orders in which marginals can be fitted:

Order 1	Order 2
a: $x(.ijk\ell)$, $x(h....)$	a: $x(.ijk\ell)$, $x(h....)$
b: $x(.ijk\ell)$, $x(hi...)$	α: $x(.ijk\ell)$, $x(h..k.)$
c: $x(.ijk\ell)$, $x(hi...)$, $x(h.j..)$	μ: $x(.ijk\ell)$, $x(h..k.)$, $x(h...\ell)$

d:	$x(.ijk\ell)$, $x(hi...)$, $x(h.j..)$, $x(h..k.)$	ν:	$x(.ijk\ell)$, $x(h..k.)$, $x(h...\ell)$, $x(hi...)$
e:	$x(.ijk\ell)$, $x(hi...)$, $x(h.j..)$, $x(h..k.)$, $x(h...\ell)$	e:	$x(.ijk\ell)$, $x(h..k.)$, $x(h...\ell)$, $x(hi...)$, $x(h.j..)$

In this case the Information Numbers $2I(x:x_a^*)$ and $2I(x:x_e^*)$ will be the same for both orders, but the intermediate effect and interaction values in the analysis of information will be different. The fit of the model selected for the analysis will depend on the set of marginals finally selected for fitting. But it is our experience, that one usually arrives at the same final model, based on a set of marginals having large effects, even following different orders, barring extreme unusual cases. Large effects for marginals tend to remain large and small effects for marginals tend to remain small independent of the sequence of fitting. See Tables 2a and 2c of Example 1 of Chapter 4 for an illustration of the foregoing. In case of doubt several different sequential fitting orders can be tried for comparison.

In the use of the computer programs to search for the large effects we note the following. To minimize the volume of paper output that must be examined and save printing time there is available a summary output option which will give only the Information Numbers for the various models in the sequence used for the search. Complete listings are called for only for those models where detailed examination seems indicated.

B. Outliers

Various procedures have been suggested for measuring the adequacy of a loglinear model with respect to specific cells in a contingency table. A cell which does not fit the model will be called an _outlier_. These outliers may lead one to reject a model which fits the other observations. In other cases, even though a model seems to fit, the outliers contribute much more than reasonable to the measure of deviation between the data and the fitted values of the model.

Many writers have used the difference between the test statistic computed for some hypothesized loglinear model and the test statistic using the same model but ignoring several preselected cells. A separate outlier computation for each cell is time-consuming. However we indicate a quick and easy approximation using MDI techniques.

The output of various computer programs we use for different hypotheses or models contains a listing called OUTLIER. The value of OUTLIER given for each cell is a lower bound for the decrease in the corresponding $2I(x:x^*)$, if that cell were not included in the estimation procedure. Large values of OUTLIER are thus those which are at least as large as very significant chi-square values for one degree of freedom. The basis for the OUTLIER computation and interpretation follows. Let x_a^* denote the MDI estimate subject to certain marginal constraints. Let x_b^* denote the MDI estimate subject to the same marginal constraints as x_a^* except that the value $x(\omega_1)$, say, is not included, so that $x_b^*(\omega_1) = x(\omega_1)$. The basic additive property of the MDI statistics states that

$$2I(x:x_a^*) = 2I(x_b^*:x_a^*) + 2I(x:x_b^*)$$

or

$$2I(x:x_a^*) - 2I(x:x_b^*) = 2I(x_b^*:x_a^*).$$

These results are summarized in the Analysis of Information Table 6.

TABLE 6

Analysis of Information

Component due to	Information	D.F.
H_a:	$2I(x:x_a^*)$	N_a
H_b: Same as H_a but	$2I(x_b^*:x_a^*)$	1
omitting $x(\omega_1)$	$2I(x:x_b^*)$	$N_b = N_a - 1$

But

$$I(x_b^*:x_a^*) = x_b^*(\omega_1) \ \ell n \frac{x_b^*(\omega_1)}{x_a^*(\omega_1)} + \sum_{\Omega - \omega_1} x_b^*(\omega) \ \ell n \frac{x_b^*(\omega)}{x_a^*(\omega)}$$

$$= x(\omega_1) \ \ell n \frac{x(\omega_1)}{x_a^*(\omega_1)} + \sum_{\Omega - \omega_1} x_b^*(\omega) \ \ell n \frac{x_b^*(\omega)}{x_a^*(\omega)} ,$$

and using the convexity property which implies that (Kullback, 1959)

$$\sum_{\Omega - \omega_1} x_b^*(\omega) \ \ell n \frac{x_b^*(\omega)}{x_a^*(\omega)} \geqq \sum_{\Omega - \omega_1} x_b^*(\omega) \ \ell n \frac{\sum_{\Omega - \omega_1} x_b^*(\omega)}{\sum_{\Omega - \omega_1} x_a^*(\omega)}$$

$$= (n - x_b^*(\omega_1)) \ \ell n \frac{n - x_b^*(\omega_1)}{n - x_a^*(\omega_1)}$$

we get from above that

$$I(x_b^*:x_a^*) \geqq x(\omega_1) \ \ell n \frac{x(\omega_1)}{x_a^*(\omega_1)} + \sum_{\Omega - \omega_1} x_b^*(\omega) \ \ell n \frac{\sum_{\Omega - \omega_1} x_b^*(\omega)}{\sum_{\Omega - \omega_1} x_a^*(\omega)}$$

$$= x(\omega_1) \ \ell n \frac{x(\omega_1)}{x_a^*(\omega_1)} + (n - x(\omega_1)) \ \ell n \frac{n - x(\omega_1)}{n - x_a^*(\omega_1)} .$$

Twice the last value can be computed and is listed as the OUTLIER entry for each cell of the complete computer output for the estimate x_a^*. The listing of outlier cells and the outlier value may also be selected as an option for the summary output. The ratio

$$\frac{2I(x:x_a^*) - 2I(x:x_b^*)}{2I(x:x_a^*)} = \frac{2I(x_b^*:x_a^*)}{2I(x:x_a^*)}$$

then indicates the percentage of the unexplained variation due to the outlier value.

Possible ways of utilizing the OUTLIER values are:

1) If more than one model seem to be contenders for the model to be used for analysis, other things being equal

(complexity, parsimonious use of parameters, etc.) choose the model with the smallest number of large OUTLIER values.

2) By omitting a small number of cells with large OUTLIER values one may find a good fit in terms of a model which is simpler and may involve fewer higher-order interaction parameters (higher-order marginals) than a more complex model including those cells.

3) Some contingency tables with ordered categories may not indicate a possible model with a good fit even using sets of higher order marginals. A study of OUTLIER values for reasonable simple models may suggest break points in the ordered categories implying a change in the values of certain interaction parameters. A partitioning of the original contingency table may then lead to a model which will provide an overall good fit with few higher order interaction parameters.

4) Sets of contingency tables indexed by some unordered categories may display over all a problem similar to that mentioned in 3) above. A grouping or partitioning as suggested by a study of OUTLIER values may be of use here too in arranging the sets of tables into more or less similar collections for appropriate model description.

Illustrations of the use of OUTLIER values as well as other aspects of the analyses are found in the examples in Chapter 4 and Chapter 8.

VIII. ALGORITHMS TO CALCULATE QUADRATIC APPROXIMATIONS

We have previously mentioned quadratic approximations to the Information Numbers as connecting links to various chi-square statistics. We now present algorithms to calculate quadratic approximations to $2I(x:x_a^*)$, $2I(x_b^*:x_a^*)$, $2I(x^*:x)$.

A. $2I(x:x_a^*)$

a) Compute x_a^*.

b) Using the T design matrix corresponding to x (including the L column), compute the matrix $\underset{\sim}{S} = \underset{\sim}{T}'\underset{\sim}{D}_a^*\underset{\sim}{T}$, where $\underset{\sim}{D}_a^*$ is a diagonal matrix whose entries are the values of x^*_a in lexicographic order.

c) Let $\underset{\sim}{S} = \begin{bmatrix} \underset{\sim}{S}_{11} & \underset{\sim}{S}_{12} \\ \underset{\sim}{S}_{21} & \underset{\sim}{S}_{22} \end{bmatrix}$, where $\underset{\sim}{S}_{11}$ is a 1 x 1 matrix,

then $\underset{\sim}{S}_{22.1} = \underset{\sim}{S}_{22} - \underset{\sim}{S}_{21}\underset{\sim}{S}_{11}^{-1}\underset{\sim}{S}_{12}$.

d) Compute $\underset{\sim}{S}_{22.1}^{-1}$.

e) Consider the marginals of x which <u>do not</u> enter into the specification of x^*_a, and let $\underset{\sim}{d}'$ be a one row matrix whose entries are the respective differences between the set of marginals just considered, in the x and x^*_a tables.

f) Let $\underset{\sim}{B}$ be that submatrix of $\underset{\sim}{S}_{22.1}^{-1}$ whose rows and columns correspond to the τ columns of the design matrix associated with the set of marginals in step e).

g) Compute $\underset{\sim}{d}'\underset{\sim}{B}\underset{\sim}{d}$, the marginals approximation to $2I(x:x^*_a)$.

h) Compute the set of natural parameters (taus) associated with the marginals considered in e) for the x distribution, and call the one row matrix of these natural parameters $\underset{\sim}{\tau}'$.

Compute $\underset{\sim}{\tau}'\underset{\sim}{B}^{-1}\underset{\sim}{\tau}$, where $\underset{\sim}{B}^{-1}$ is the inverse of the matrix $\underset{\sim}{B}$ in f), this is the natural parameter (tau) approximation to $2I(x:x^*_a)$.

i) The marginals approximation is also equal to

$$\Sigma\ (x - x^*_a)^2/x^*_a$$

B. $2\,I(x^*_b:x^*_a)$

a) Compute x^*_b, x^*_a.

b) Using the T design matrix corresponding to x^*_b (including the L column), compute the matrix $\underset{\sim}{S} = \underset{\sim}{T}'\underset{\sim}{D}^*_a\underset{\sim}{T}$, where $\underset{\sim}{D}^*_a$ is a diagonal matrix whose entries are the values of x^*_a in lexicographic order.

c) Let $\underset{\sim}{S} = \begin{bmatrix} \underset{\sim}{S}_{11} & \underset{\sim}{S}_{12} \\ \underset{\sim}{S}_{21} & \underset{\sim}{S}_{22} \end{bmatrix}$, where $\underset{\sim}{S}_{11}$ is a 1 x 1 matrix, then

$\underset{\sim}{S}_{22.1} = \underset{\sim}{S}_{22} - \underset{\sim}{S}_{21}\underset{\sim}{S}_{11}^{-1}\underset{\sim}{S}_{12}$.

d) Compute $\underset{\sim}{S}_{22.1}^{-1}$.

e) Consider the marginals which enter into the specification of x_b^* but <u>not</u> in x_a^* and let $\underset{\sim}{d}'$ be a one row matrix whose entries are the respective differences between the set of marginals just considered in the x_b^* and x_a^* tables.

f) Let $\underset{\sim}{B}$ be that submatrix of $\underset{\sim}{S}_{22.1}^{-1}$ whose rows and columns correspond to the natural parameter (tau) columns of the design matrix associated with the set of marginals in step e).

g) Compute $\underset{\sim}{d}'\underset{\sim}{B}\underset{\sim}{d}$, the marginals approximation to $2I(x_b^*:x_a^*)$.

h) Compute the set of natural parameters (taus) associated with the marginals considered in e) for the x_b^* distribution and call the one row matrix of these natural parameters $\underset{\sim}{\tau}'$.

Compute $\underset{\sim}{\tau}'\underset{\sim}{B}^{-1}\underset{\sim}{\tau}$ where $\underset{\sim}{B}^{-1}$ is the inverse of the matrix $\underset{\sim}{B}$ in f), this is the natural parameter (tau) approximation to $2I(x_b^*:x_a^*)$.

i) The marginals approximation is also equal to

$$\Sigma\ (x_b^* - x_a^*)^2/x_a^*.$$

C. 2I(x*:x), ECP

a) Using the T design matrix corresponding to x* (including the L column), compute the matrix $\underset{\sim}{S} = \underset{\sim}{T}'\underset{\sim}{D}_x\underset{\sim}{T}$, where $\underset{\sim}{D}_x$ is a diagonal matrix whose entries are the values of x in lexicographic order.

b) Let $\underset{\sim}{S} = \begin{bmatrix} \underset{\sim}{S}_{11} & \underset{\sim}{S}_{12} \\ \underset{\sim}{S}_{21} & \underset{\sim}{S}_{22} \end{bmatrix}$, where $\underset{\sim}{S}_{11}$ is a 1 x 1 matrix, then

$$\underset{\sim}{S}_{22.1} = \underset{\sim}{S}_{22} - \underset{\sim}{S}_{21}\underset{\sim}{S}_{11}\underset{\sim}{S}_{12}.$$

c) Compute $\underset{\sim}{S}_{22.1}^{-1}$.

d) Let $\underset{\sim}{d}'$ be a one row matrix whose entries are the respective differences between the $\sum_{\omega} T_i(\omega)x^*(\omega)$ and $\sum_{\omega} T_i(\omega)x(\omega)$. In the case when $x^*(\omega)$ is specified by conditions external to the observed values, ECP, the value of $\sum_{\omega} T_i(\omega)x^*(\omega)$ is specified without having to compute $x^*(\omega)$.

e) Compute $\underset{\sim}{d}'\underset{\sim}{S}_{22.1}^{-1}\underset{\sim}{d}$, the moment approximation to $2I(x^*:x)$. Note that this can be obtained _without_ computing x^*.

f) The approximation $\Sigma\ (x^* - x)^2/x$ requires the prior computation of x^*.

IX. CONFIDENCE INTERVALS

One can determine asymptotic _simultaneous sets of confidence intervals_ for a set of the estimates of the natural parameters (taus), or for their linear combinations, in the model selected for detailed analysis. In addition to the natural parameter (tau) estimates, the computer outputs will also provide the asymptotic covariance matrix of the estimates. The procedure depends on the following argument in matrix notation (Ferguson, 1967, page 282; Kullback, 1959, page 196).

Let $x_i = X_i - \mu_i$, $i = 1, 2, \ldots, n$, where μ_i is the expected value of X_i, and let $\underset{\sim}{\Sigma}$ be the covariance matrix of the x_i. Set $\underset{\sim}{x}' = (x_1, x_2, \ldots, x_n)$, $\underset{\sim}{a}' = (a_1, a_2, \ldots, a_n)$, where the a's are any real numbers, then $\underset{\sim}{a}'\ \underset{\sim}{x}\ \underset{\sim}{x}'\ \underset{\sim}{a}/\underset{\sim}{a}'\ \underset{\sim}{\Sigma}\ \underset{\sim}{a} \leqq \underset{\sim}{x}'\ \underset{\sim}{\Sigma}^{-1}\ \underset{\sim}{x}$. If the x's are normally distributed, $\underset{\sim}{x}'\underset{\sim}{\Sigma}^{-1}\underset{\sim}{x}$ has a chi-square distribution with n degrees of freedom, and Prob $(\underset{\sim}{a}'\ \underset{\sim}{x}\ \underset{\sim}{x}'\ \underset{\sim}{a}/\underset{\sim}{a}'\ \underset{\sim}{\Sigma}\ \underset{\sim}{a} \leqq \chi^2_{n,\alpha}$ for all a) $\geqq 1 - \alpha$.

Thus in particular for $\underset{\sim}{a}' = (1, 0, \ldots, 0)$, $\underset{\sim}{a}' = (0, 1, 0, \ldots, 0)$, $\underset{\sim}{a}' = (0, 0, 1, 0, \ldots, 0)$, ... , $\underset{\sim}{a}' = (0, 0, \ldots, 0, 1)$, there follows the joint set of inequalities with probability $\geqq 1 - \alpha$,

$$x_1^2 \leqq \sigma_1^2 \chi_{n,\alpha}^2$$
$$x_2^2 \leqq \sigma_2^2 \chi_{n,\alpha}^2$$
$$\vdots \qquad \vdots$$
$$x_n^2 \leqq \sigma_n^2 \chi_{n,\alpha}^2$$

The set of inequalities is converted into the set of simultaneous confidence intervals with confidence coefficient $1 - \alpha$,

$$X_1 - (\sigma_1^2 \chi_{n,\alpha}^2)^{1/2} \leqq \mu_1 \leqq X_1 + (\sigma_1^2 \chi_{n,\alpha}^2)^{1/2}$$
$$X_2 - (\sigma_2^2 \chi_{n,\alpha}^2)^{1/2} \leqq \mu_2 \leqq X_2 + (\sigma_2^2 \chi_{n,\alpha}^2)^{1/2}$$
$$\vdots \qquad \vdots \qquad \vdots$$
$$X_n - (\sigma_n^2 \chi_{n,\alpha}^2)^{1/2} \leqq \mu_n \leqq X_n + (\sigma_n^2 \chi_{n,\alpha}^2)^{1/2}$$

We shall use this procedure in some of the examples in Chapter 4 and Chapter 8.

Chapter 4

APPLICATIONS

In this chapter we shall consider nine examples illustrating various aspects of the ICP methodology applied to the analysis of real data. Included in the examples are two-way, three-way, four-way, and five-way contingency tables.

The analyses discuss which model is appropriate as the initial or basic distribution, and consider the sequential search for large effects which would suggest the final model for analysis. A test for the equality of certain natural parameters (taus) is illustrated, as well as the computation of the MDI estimate satisfying the equality constraint. Another example shows how the inheritance property is used to compute an MDI estimate having some natural parameters (taus) predetermined. A large-sample estimate of the covariance matrix of the estimates of the natural parameters (taus) of special interest in certain of the examples is given, and simultaneous sets of confidence intervals for these natural parameters (taus) computed. Illustrations of the use of OUTLIER both with ordered and unordered categories are given. The use of a logit representation is shown, as well as log-odds and a multiplicative model for the odds. The concept of no second-order or three-factor interaction is discussed as well as the interpretation of association and interaction. In most of the examples the dependent classification is dichotomous but there is also included an example with a polychotomous dependent classification. It is pointed out that the model determines the logit representation uniquely but the logit representation does not necessarily imply the model uniquely.

For the sake of simplicity in writing, we shall often suppress the distinction between the natural parameters of the estimates denoting them by the symbol τ with appropriate superscripts and subscripts. Similarly by the phrase covariance matrix of the estimates of the natural parameters (taus) we shall mean a large sample estimate of such a covariance matrix. The reader should bear in mind that quantities calculated by using the observed values are estimates of the corresponding parameters.

EXAMPLE 1

Classification of multivariate dichotomous populations. This example illustrates the analysis of a five-way 2 x 2 x 2 x 2 x 2 contingency table. It introduces the use of log-odds or logit representation, and the multiplicative version of the odds as a product of factors. It also illustrates the interpretation of the natural parameters, and the effect of interaction on the numerical value of the association between classifications. It considers several models with respect to the marginals fitted, the design matrices, and the detailed hierarchical analysis of information. For one of the models, the covariance matrix of the estimates of the natural parameters (taus) is given as well as a simultaneous set of 95% confidence intervals.

MULTIWAY CONTINGENCY TABLE ANALYSIS APPLIED TO THE CLASSIFICATION OF MULTIVARIATE DICHOTOMOUS POPULATIONS

Introduction

Multiway contingency tables, or cross-classifications of vectors of discrete random variables, provide a useful approach to the analysis of multivariate discrete data. In the particular application we consider, the individual variates are dichotomous or binary. Note however that the procedures and analysis are not restricted to dichotomous or binary data but are also applicable to polychotomous variates.

For background on the study and problem leading to the data we consider see Solomon (1960). In Ku and Kullback (1969) MDI estimation procedures were applied to problems of multivariate binary data in information systems, such as communication, pattern recognition, and learning systems. In Cox (1972) there is a review of methods and models for the analysis of multivariate binary data and Solomon's data is given as a typical example. Martin and Bradley (1972) developed a model based on a set of orthogonal polynomials and applied it to Solomon's data. We remark that our procedure based on the principle of MDI estimation applied to the analysis of multiway contingency tables yields a result practically equivalent to that of Martin and Bradley (1972). Goodman (1973) discusses Solomon's data in relation to methods for selecting models for contingency tables.

Solomon's Data

A total of 2982 high-school seniors were given an attitude questionnaire to assess their attitude towards science. The students were also classified on the basis of an IQ test into high IQ, the upper half, and low IQ, the lower half. The sixteen possible response vectors to each of four agree-disagree responses were tabulated. The problem of interest was to determine whether the response vectors could be used as a basis for classifying the students into one of two classes and evaluate possible classification procedures.

Contingency Table Analysis

We shall treat the data given in Table 1 as a five way 2 x 2 x 2 x 2 x 2 contingency table, denoting the original observations by $x(hijk\ell)$, where

Characteristic	Index	1	2
IQ	h	low IQ	high IQ
Response 1	i	disagree	agree
Response 2	j	disagree	agree
Response 3	k	disagree	agree
Response 4	ℓ	disagree	agree

As a first overview of the data to determine the marginals and their related natural parameters which may furnish significant values in the loglinear representation of the MDI estimates, we list in Table 2a, Analysis of Information, a sequential hierarchical study of interaction and effect type measures.

The procedure of analysis can be described in conjunction with Table 2a, as follows:

We start with the null hypothesis that the IQ groupings are homogeneous over the sixteen response vectors (Kullback, 1959, Chap. 8). The MDI estimate is of the form $x_a^*(hijk\ell) = x(h....)x(.ijk\ell)/n$ with

$$2I(x:x_a^*) = 2\Sigma\Sigma\Sigma\Sigma\Sigma x(hijk\ell)\ell n\ x(hijk\ell)n/x(h....)x(.ijk\ell)$$

which equals 68.369 with 15 degrees of freedom. Note that this is equivalent to the test for homogeneity in a 2 x 16 table. This null hypothesis is rejected and the subsequent study of effect and interaction type measures is an attempt to find a good fit to the data and account for much of the total variation as measured by $2I(x:x_a^*)$.

Note that we have imposed no a priori structure on the presence or absence of association among the four questions.

Thus, for example, if it were known beforehand that responses to questions 1 and 2 jointly, were independent of joint responses to questions 3 and 4, we would fit marginals $x(h....)$, $x(.ij..)$ and $x(...k\ell)$ instead of $x(h....)$ and $x(.ijk\ell)$. But, in the absence of such knowledge, it is appropriate to use $x(.ijk\ell)$ to retain the association structure inherent in the data.

The next estimate x_b^* includes the marginal $x(hi...)$ and corresponds to the model that IQ is homogeneous over the response to the last three questions, given the response to the first question (Kullback, 1959, Chap. 8). Thus

$$x_b^*(hijk\ell) = x(hi...)x(.ijk\ell)/x(.i...)$$

with

$$2I(x:x_b^*) = 2\Sigma\Sigma\Sigma\Sigma\Sigma x(hijk\ell) \ln x(hijk\ell)x(.i...)/x(hi...)x(.ijk\ell),$$

and

$$2I(x_b^*:x_a^*) = 2\Sigma\Sigma x(hi...) \ln x(hi...)n/x(h....)x(.i...).$$

The latter, a measure of the effect of the natural parameters associated with $x(hi...)$, tests the null hypothesis that IQ is homogeneous over the response to the first question. The calculated value of $2I(x_b^*:x_a^*)$ is 2.376 with 1 D.F., which is not significant, but $2I(x:x_b^*) = 65.993$ with 14 D.F. is highly significant.

The MDI estimate x_c^* is calculated by fitting one more two-way marginal $x(h.j..)$. In this case x_c^* cannot be expressed explicitly as a product of marginals and must be obtained by iteration.

TABLE 1

Solomon's Data-Classification Procedures

$ijk\ell$	Observed Low IQ $x(1ijk\ell)$	Martin & Bradley	Estimates $x_e^*(1ijk\ell)$	$x_v^*(1ijk\ell)$	$x_w^*(1ijk\ell)$
2222	62	74.56	74.589	76.097	70.156
2221	70	67.30	67.296	66.198	71.600
2212	31	31.32	31.329	31.943	29.827
2211	41	37.74	37.780	37.337	39.884
2122	283	266.76	266.570	271.120	275.979
2121	253	259.17	259.322	254.876	250.769
2112	200	193.45	193.625	196.841	200.037
2111	305	314.50	314.491	310.589	306.748
1222	14	12.10	12.156	10.866	9.914
1221	11	9.20	9.182	9.929	10.760
1212	11	9.68	9.659	8.776	8.102
1211	14	12.02	12.010	12.855	12.756
1122	31	33.63	33.623	30.125	30.820
1121	46	47.37	47.263	50.789	50.001
1112	37	47.54	47.450	43.233	44.163
1111	82	74.67	74.656	79.426	78.482
	1491				

Observed High IQ $x(2ijk\ell)$	Martin & Bradley	$x_e^*(2ijk\ell)$	Estimates $x_v^*(2ijk\ell)$	$x_w^*(2ijk\ell)$
122	109.45	109.414	107.904	113.844
68	70.71	70.703	71.802	66.400
33	32.68	32.671	32.057	34.173
25	28.26	28.219	28.662	26.115
329	345.24	345.429	340.879	336.020
247	240.83	240.679	245.125	249.232
172	178.55	178.376	175.160	171.963
217	207.50	207.508	211.411	215.252
20	21.90	21.844	23.135	24.085
10	11.80	11.818	11.071	10.240
11	12.32	12.341	13.224	13.898
9	10.98	10.990	10.144	9.244
56	53.37	53.375	56.874	56.179
55	53.63	53.737	50.211	50.999
64	53.46	53.550	57.767	56.837
53	60.33	60.346	55.574	56.517
1491				

Again, $2I(x_c^*:x_b^*)$, a measure of the effect of the additional natural parameters associated with x(h.j..), is marginally significant at the 5% level (1 D.F.) and $2I(x:x_c^*) = 61.728$ is significant (13 D.F.). One more two-way marginal x(h..k.) is then fitted, which still gives a significant value for $2I(x:x_d^*)$. The MDI estimate $x_e^*(hijk\ell)$, whose values are in Table 1, was selected because $2I(x:x_e^*) = 16.307$ (11 D.F.) implies an acceptable fit, the estimate is symmetric with respect to the four statements and is comparable to the first order model estimate of Martin and Bradley (1972), whose values are also listed in Table 1.

From the design matrix or loglinear representation in Table 5, we obtain the parametric representation for the log-odds (low IQ/high IQ) $\ell n(x_e^*(1ijk\ell)/x_e^*(2ijk\ell))$ over the sixteen response vectors as given in Table 3a. Thus, for example

$$\ell n\ x_e^*(11111)/x_e^*(21111) = \tau_1^h + \tau_{11}^{hi} + \tau_{11}^{hj} + \tau_{11}^{hk} + \tau_{11}^{h\ell},$$

that is, a linear regression of the log-odds in terms of an overall average τ_1^h and the main effects of each component of the response vector, namely, τ_{11}^{hi}, τ_{11}^{hj}, τ_{11}^{hk}, $\tau_{11}^{h\ell}$. The numerical values of the log-odds and the natural parameters are easily obtained from the estimated cell entries and are given in Table 3a. It is clear that the odds may be expressed as a multiplicative model. The odds and the odds factors are easier to interpret than log-odds. From the log-odds representation above we find

$$x_e^*(11111)/x_e^*(21111) = \exp(\tau_1^h)\ \exp(\tau_{11}^{hi})\ \exp(\tau_{11}^{hj})\ \exp(\tau_{11}^{hk})\ \exp(\tau_{11}^{h\ell})$$

and from the values in Table 3a have

$$1.237 = (.682)(.816)(1.132)(1.406)(1.396).$$

In Table 3d we present the odds-factors **table** for x_e^*.

We note from Table 3a that

$$\ell n\ x_e^*(1ijk1)/x_e^*(2ijk1) - \ell n\ x_e^*(1ijk2)/x_e^*(2ijk2) = \tau_{11}^{h\ell} = 0.3338,$$

that is, a change from disagree to agree on the fourth statement is associated with an increase of 0.3338 in the log-odds (low IQ/high IQ). Note also that $\tau_{11}^{h\ell}$ represents the association between IQ and response to the fourth statement as measured by the log cross-product ratio (log relative odds)

$$\tau_{11}^{h\ell} = \ell n\ x_e^*(1ijk1)x_e^*(2ijk2)/x_e^*(2ijk1)x_e^*(1ijk2),$$

and is the same for all eight levels of the responses to statements one, two and three.

Similarly, it is found that

$$\ell n\ x_e^*(1ij1\ell)/x_e^*(2ij1\ell) - \ell n\ x_e^*(1ij2\ell)/x_e^*(2ij2\ell) = \tau_{11}^{hk} = 0.3411,$$

$$\ell n\ x_e^*(1i1k\ell)/x_e^*(2i1k\ell) - \ell n\ x_e^*(1i2k\ell)/x_e^*(2i2k\ell) = \tau_{11}^{hj} = 0.1240,$$

$$\ell n\ x_e^*(11jk\ell)/x_e^*(21jk\ell) - \ell n\ x_e^*(12jk\ell)/x_e^*(22jk\ell) = \tau_{11}^{hi} = -\ 0.2030.$$

The covariance matrix of the natural parameters (taus) given above for x_e^* is given in Table 6. A simultaneous set of 95% confidence intervals for τ_{11}^{hi}, τ_{11}^{hj}, τ_{11}^{hk}, $\tau_{11}^{h\ell}$ based on $\tau \pm (9.488\ \mathrm{Var}\ \tau)^{1/2}$ and the variances in Table 6 is given by:

τ_{11}^{hi}: - 0.5039, 0.0979 $\qquad$ τ_{11}^{hk}: 0.1081, 0.5743

τ_{11}^{hj}: - 0.1715, 0.4195 $\qquad$ $\tau_{11}^{h\ell}$: 0.1047, 0.5629

Classification

Since $x(1....) = x_e^*(1....) = 1491$, and $x(2....) = x_e^*(2....) = 1491$, we assign a response vector $(ijk\ell)$ to the region

E_1: classify as population h = 1 (low IQ), when

$$\ell n\ x_e^*(1ijk\ell)/x_e^*(2ijk\ell) \geq 0$$

and to the complementary region

E_2: classify as population h = 2 (high IQ), when

$$\ell n\ x_e^*(1ijk\ell)/x_e^*(2ijk\ell) < 0.$$

If we set

$$\mu_1(E_1) = \sum_{(ijk\ell)\varepsilon E_1} \frac{x_e^*(1ijk\ell)}{1491}, \quad \mu_2(E_1) = \sum_{(ijk\ell)\varepsilon E_1} \frac{x_e^*(2ijk\ell)}{1491},$$

then the probability of error of the classification procedure is (Kullback, 1959, pp. 4, 69, 80),

$$\text{Prob Error} = p\mu_2(E_1) + q\mu_1(E_2) = (\mu_2(E_1) + \mu_1(E_2))/2$$

since here $p = x(2....)/2982 = 1/2$, $q = x(1....)/2982 = 1/2$.

The relevant computations with $x_e^*(hijk\ell)$ are given in Table 4b and show that the Prob. Error = 0.444. The corresponding computations with the original data $x(hijk\ell)$ are given in Table 4a and yield Prob. Error = 0.441.

Other Estimates

From Table 2a we see that $2I(x_m^*:x_g^*) = 4.316$ with 1 D.F. This shows that inclusion of the marginals $x(hi..\ell)$ in addition to those corresponding to x^*_g may be of importance. Since $2I(x^*_p:x^*_n) =$ 3.181 with 1 D.F. a similar remark applies to the marginals $x(h.j.\ell)$ in relation to marginals corresponding to x^*_n. The MDI estimate x^*_v excludes the three-way marginals $x(hij..)$ and $x(hi.k.)$ used in x^*_g but uses $x(hi..\ell)$ in addition to the marginals in x^*_e. The MDI estimate x^*_w uses the marginals $x(h.j.\ell)$ in addition to those in x^*_v. The estimates are given in Table 1 and the relevant analysis of information given in Table 2b.

The values of the log-odds, parametric representation, and the associated natural parameters are given in Table 3b for $x^*_v(hijk\ell)$ and in Table 3c for $x^*_w(hijk\ell)$. Note from Table 3b that

$$\ell n \frac{x^*_v(11jk1)}{x^*_v(21jk1)} - \ell n \frac{x^*_v(11jk2)}{x^*_v(21jk2)} = \tau^{h\ell}_{11} + \tau^{hi\ell}_{111} = 0.6469,$$

$$\ell n \frac{x^*_v(12jk1)}{x^*_v(22jk1)} - \ell n \frac{x^*_v(12jk2)}{x^*_v(22jk2)} = \tau^{h\ell}_{11} = 0.2680,$$

$$\ell n \frac{x^*_v(11jk1)}{x^*_v(21jk1)} - \ell n \frac{x^*_v(12jk1)}{x^*_v(22jk1)} = \tau^{hi}_{11} + \tau^{hi\ell}_{111} = -\ 0.0276,$$

$$\ell n \frac{x^*_v(11jk2)}{x^*_v(21jk2)} - \ell n \frac{x^*_v(12jk2)}{x^*_v(22jk2)} = \tau^{hi}_{11} = -\ 0.4065.$$

From Table 3c, it is found for example, that

$$\ell n \frac{x^*_w(111k1)}{x^*_w(211k1)} - \ell n \frac{x^*_w(111k2)}{x^*_w(211k2)} = \tau^{h\ell}_{11} + \tau^{hi\ell}_{111} + \tau^{hj\ell}_{111} = 0.5806,$$

$$\ell n \frac{x^*_w(121k1)}{x^*_w(221k1)} - \ell n \frac{x^*_w(121k2)}{x^*_w(221k2)} = \tau^{h\ell}_{11} + \tau^{hj\ell}_{111} = 0.2030,$$

$$\ell n \frac{x_w^*(112k1)}{x_w^*(212k1)} - \ell n \frac{x_w^*(112k2)}{x_w^*(212k2)} = \tau_{11}^{h\ell} + \tau_{111}^{hi\ell} = 0.9371,$$

$$\ell n \frac{x_w^*(122k1)}{x_w^*(222k1)} - \ell n \frac{x_w^*(122k2)}{x_w^*(222k2)} = \tau_{11}^{h\ell} = 0.5595,$$

reflecting the interactions of the responses to the first and fourth statements respectively first, second and fourth statements.

The computation of the probability of error using the estimates $x_v^*(hijk\ell)$ and $x_w^*(hijk\ell)$ is shown in Table 4c and 4d respectively, and yields probabilities of error 0.444 and 0.446.

Remark

Martin and Bradley (1972) examined Solomon's data in terms of an estimate they called a first-order or linear model. These estimated values are given in Table 1. It turns out that although the underlying approaches are different, the Martin and Bradley parameters, their a_i, and estimates are practically the same as those for $x_e^*(hijk\ell)$. From Martin and Bradley (1972, pp. 216-217) we note that

$$\ell n \frac{x_e^*(12222)}{x_e^*(22222)} = \tau_1^h = \ell n \frac{1 + a_0 + a_1 + a_2 + a_3 + a_4}{1 - a_0 - a_1 - a_2 - a_3 - a_4},$$

$$\ell n \frac{x_e^*(12221)}{x_e^*(22221)} = \tau_1^h + \tau_{11}^{h\ell} = \ell n \frac{1 + a_0 + a_1 + a_2 + a_3 - a_4}{1 - a_0 - a_1 - a_2 - a_3 + a_4},$$

$$\ell n \frac{x_e^*(12212)}{x_e^*(22212)} = \tau_1^h + \tau_{11}^{hk} = \ell n \frac{1 + a_0 + a_1 + a_2 - a_3 + a_4}{1 - a_0 - a_1 - a_2 + a_3 - a_4},$$

$$\ln \frac{x_e^*(12122)}{x_e^*(22122)} = \tau_1^h + \tau_{11}^{hj} = \ln \frac{1 + a_0 + a_1 - a_2 + a_3 + a_4}{1 - a_0 - a_1 + a_2 - a_3 - a_4},$$

$$\ln \frac{x_e^*(11222)}{x_e^*(21222)} = \tau_1^h + \tau_{11}^{hi} = \ln \frac{1 + a_0 - a_1 + a_2 + a_3 + a_4}{1 - a_0 + a_1 - a_2 - a_3 - a_4},$$

or to a first approximation of the logarithm

$$\tau_1^h = 2a_0 + 2a_1 + 2a_2 + 2a_3 + 2a_4,$$

$$\tau_1^h + \tau_{11}^{h\ell} = 2a_0 + 2a_1 + 2a_2 + 2a_3 - 2a_4,$$

$$\tau_1^h + \tau_{11}^{hk} = 2a_0 + 2a_1 + 2a_2 - 2a_3 + 2a_4,$$

$$\tau_1^h + \tau_{11}^{hj} = 2a_0 + 2a_1 - 2a_2 + 2a_3 + 2a_4,$$

$$\tau_1^h + \tau_{11}^{hi} = 2a_0 - 2a_1 + 2a_2 + 2a_3 + 2a_4.$$

It is found that

$$\tau_{11}^{h\ell} = -4a_4,$$

$$\tau_{11}^{hk} = -4a_3,$$

$$\tau_{11}^{hj} = -4a_2,$$

$$\tau_{11}^{hi} = -4a_1.$$

The values of the parameters given by Martin and Bradley (1972,

Table 3, p. 217) are $a_0 = -0.042$, $a_1 = 0.049$, $a_2 = -0.031$, $a_3 = -0.084$, $a_4 = -0.082$ so that

$$\tau_{11}^{h\ell} = 0.3338 = 0.334, \quad -4a_4 = 0.328,$$

$$\tau_{11}^{hk} = 0.3411 = 0.341, \quad -4a_3 = 0.336,$$

$$\tau_{11}^{hj} = 0.1240 = 0.124, \quad -4a_2 = 0.124,$$

$$\tau_{11}^{hi} = -0.2030 = -0.203, \quad -4a_1 = -0.196.$$

The computation for the probability of error using the estimates are shown in Table 4e and yields a probability of error 0.445. (Martin and Bradley give a value of the risk as 0.455).

It seems reasonable to conclude that the four statements do not provide a strong discrimination. This is not surprising in view of the fact that the ratio of largest to smallest odds is approximately 3 to 1.

The Analysis of Information Table 2c shows the results for a different ordering of the first five estimates in Table 2a. In this regard see the discussion in Chapter 3 Section VII A.

Note that the effect of $x(h..k.)$ in Table 2a is 25.230 and in Table 2c it is 25.615; the effect of $x(h...\ell)$ is 20.191 in Table 2a and 20.350 in Table 2c. The effects of $x(hi...)$ and $x(h.j..)$ are respectively 2.376 and 4.265 in Table 2a and 4.424 and 1.673 in Table 2c.

The response to the second statement does not seem to play a significant role and one might be inclined to analyze the estimate x_ν^*. We leave this as an exercise for the reader.

TABLE 2a

Analysis of Information

	Marginals Fitted	Information	D.F.
a)	x(.ijkℓ), x(h....)	$2I(x:x_a^*) = 68.369$	15
b)	x(.ijkℓ), x(hi...)	$2I(x_b^*:x_a^*) = 2.376$	1
		$2I(x:x_b^*) = 65.993$	14
c)	x(.ijkℓ), x(hi...), x(h.j..)	$2I(x_c^*:x_b^*) = 4.265$	1
		$2I(x:x_c^*) = 61.728$	13
d)	x(.ijkℓ), x(hi...), x(h.j..),	$2I(x_d^*:x_c^*) = 25.230$	1
	x(h..k.)	$2I(x:x_d^*) = 36.498$	12
e)	x(.ijkℓ), x(hi...), x(h.j..),	$2I(x_e^*:x_d^*) = 20.191$	1
	x(h..k.), x(h...ℓ)	$2I(x:x_e^*) = 16.307$	11
f)	x(.ijkℓ), x(h..k.), x(h...ℓ),	$2I(x_f^*:x_e^*) = 3.016$	1
	x(hij..)	$2I(x:x_f^*) = 13.291$	10
g)	x(.ijkℓ), x(h...ℓ), x(hij..),	$2I(x_g^*:x_f^*) = 0.042$	1
	x(hi.k.)	$2I(x:x_g^*) = 13.249$	9
m)	x(.ijkℓ), x(hij..), x(hi.k.),	$2I(x_m^*:x_g^*) = 4.316$	1
	x(hi..ℓ)	$2I(x:x_m^*) = 8.933$	8
n)	x(.ijkℓ), x(hij..), x(hi.k.),	$2I(x_n^*:x_m^*) = 0.983$	1
	x(hi..ℓ), x(h.jk.)	$2I(x:x_n^*) = 7.950$	7
p)	x(.ijkℓ), x(hij..), x(hi.k.),	$2I(x_p^*:x_n^*) = 3.181$	1
	x(hi..ℓ), x(h.jk.), x(h.j.ℓ)	$2I(x:x_p^*) = 4.769$	6

Analysis of Information (continued)

	Marginals Fitted	Information	D.F.
q)	x(.ijkℓ), x(hij..), x(hi.k.),	$2I(x_q^*:x_p^*) = 0.219$	1
	x(hi..ℓ), x(h.jk.), x(h.j.ℓ),	$2I(x:x_q^*) = 4.550$	5
	x(h..kℓ)		
r)	x(.ijkℓ), x(hi..ℓ), x(h.j.ℓ),	$2I(x_r^*:x_q^*) = 0.346$	1
	x(h..kℓ), x(hijk.)	$2I(x:x_r^*) = 4.204$	4
s)	x(.ijkℓ), x(h..kℓ), x(hijk.),	$2I(x_s^*:x_r^*) = 2.303$	1
	x(hij.ℓ)	$2I(x:x_s^*) = 1.901$	3
t)	x(.ijkℓ), x(hijk.), x(hij.ℓ),	$2I(x_t^*:x_s^*) = 1.375$	1
	x(hi.kℓ)	$2I(x:x_t^*) = 0.526$	2
u)	x(.ijkℓ), x(hijk.), x(hij.ℓ),	$2I(x_u^*:x_t^*) = 0.361$	1
	x(hi.kℓ), x(h.jkℓ)	$2I(x:x_u^*) = 0.165$	1

TABLE 2b

Analysis of Information

	Marginals Fitted	Information	D.F.
e)	x(.ijkℓ), x(hi...), x(h.j..),	$2I(x:x_e^*) = 16.307$	11
	x(h..k.), x(h...ℓ)		
v)	x(.ijkℓ), x(h.j..), x(h..k.),	$2I(x_v^*:x_e^*) = 3.735$	1
	x(hi..ℓ)	$2I(x:x_v^*) = 12.572$	10
w)	x(.ijkℓ), x(h..k.), x(hi..ℓ),	$2I(x_w^*:x_v^*) = 3.443$	1
	x(h.j.ℓ)	$2I(x:x_w^*) = 9.129$	9

TABLE 2c

Analysis of Information

	Marginals Fitted	Information	D.F.
a)	$x(.ijk\ell)$, $x(h....)$	$2I(x:x^*_a) = 68.369$	15
α)	$x(.ijk\ell)$, $x(h..k.)$	$2I(x^*_\alpha:x^*_a) = 25.615$	1
		$2I(x:x^*_\alpha) = 42.754$	14
μ)	$x(.ijk\ell)$, $x(h..k.)$, $x(h...\ell)$	$2I(x^*_\mu:x^*_\alpha) = 20.350$	1
		$2I(x:x^*_\mu) = 22.404$	13
ν)	$x(.ijk\ell)$, $x(h..k.)$, $x(h...\ell)$,	$2I(x^*_\nu:x^*_\mu) = 4.424$	1
	$x(hi...)$	$2I(x:x^*_\nu) = 17.980$	12
e)	$x(.ijk\ell)$, $x(h..k.)$, $x(h...\ell)$,	$2I(x^*_e:x^*_\nu) = 1.673$	1
	$x(hi...)$, $x(h.j..)$	$2I(x:x^*_e) = 16.307$	11

TABLE 3a

Log-odds $\ln x_e^*(1ijk\ell)/x_e^*(2ijk\ell)$

$ijk\ell$	Parametric representation					log-odds
1111	τ_1^h	$+\tau_{11}^{hi}$	$+\tau_{11}^{hj}$	$+\tau_{11}^{hk}$	$+\tau_{11}^{h\ell}$	0.2128
1112	τ_1^h	$+\tau_{11}^{hi}$	$+\tau_{11}^{hj}$	$+\tau_{11}^{hk}$		-0.1210
1121	τ_1^h	$+\tau_{11}^{hi}$	$+\tau_{11}^{hj}$		$+\tau_{11}^{h\ell}$	-0.1284
1122	τ_1^h	$+\tau_{11}^{hi}$	$+\tau_{11}^{hj}$			-0.4621
1211	τ_1^h	$+\tau_{11}^{hi}$		$+\tau_{11}^{hk}$	$+\tau_{11}^{h\ell}$	0.0888
1212	τ_1^h	$+\tau_{11}^{hi}$		$+\tau_{11}^{hk}$		-0.2450
1221	τ_1^h	$+\tau_{11}^{hi}$			$+\tau_{11}^{h\ell}$	-0.2524
1222	τ_1^h	$+\tau_{11}^{hi}$				-0.5861
2111	τ_1^h		$+\tau_{11}^{hj}$	$+\tau_{11}^{hk}$	$+\tau_{11}^{h\ell}$	0.4158
2112	τ_1^h		$+\tau_{11}^{hj}$	$+\tau_{11}^{hk}$		0.0820
2121	τ_1^h		$+\tau_{11}^{hj}$		$+\tau_{11}^{h\ell}$	0.0746
2122	τ_1^h		$+\tau_{11}^{hj}$			-0.2592
2211	τ_1^h			$+\tau_{11}^{hk}$	$+\tau_{11}^{h\ell}$	0.2918
2212	τ_1^h			$+\tau_{11}^{hk}$		-0.0420
2221	τ_1^h				$+\tau_{11}^{h\ell}$	-0.0494
2222	τ_1^h					-0.3831

$\tau_1^h = -0.3831,\ \tau_{11}^{hi} = -0.2030,\ \tau_{11}^{hj} = 0.1240$

$\tau_{11}^{hk} = 0.3411,\ \tau_{11}^{h\ell} = 0.3338$

TABLE 3b

Log-odds $\ell n\ x^*_v(1ij k\ell)/x^*_v(2ijk\ell)$

$ijk\ell$	Parametric representation						log-odds
1111	τ_1^h	$+\tau_{11}^{hi}$	$+\tau_{11}^{hj}$	$+\tau_{11}^{hk}$	$+\tau_{11}^{h\ell}$	$+\tau_{111}^{hi\ell}$	0.3571
1112	τ_1^h	$+\tau_{11}^{hi}$	$+\tau_{11}^{hj}$	$+\tau_{11}^{hk}$			-0.2898
1121	τ_1^h	$+\tau_{11}^{hi}$	$+\tau_{11}^{hj}$		$+\tau_{11}^{h\ell}$	$+\tau_{111}^{hi\ell}$	0.0115
1122	τ_1^h	$+\tau_{11}^{hi}$	$+\tau_{11}^{hj}$				-0.6355
1211	τ_1^h	$+\tau_{11}^{hi}$		$+\tau_{11}^{hk}$	$+\tau_{11}^{h\ell}$	$+\tau_{111}^{hi\ell}$	0.2366
1212	τ_1^h	$+\tau_{11}^{hi}$		$+\tau_{11}^{hk}$			-0.4101
1221	τ_1^h	$+\tau_{11}^{hi}$			$+\tau_{11}^{h\ell}$	$+\tau_{111}^{hi\ell}$	-0.1088
1222	τ_1^h	$+\tau_{11}^{hi}$					-0.7557
2111	τ_1^h		$+\tau_{11}^{hj}$	$+\tau_{11}^{hk}$	$+\tau_{11}^{h\ell}$		0.3847
2112	τ_1^h		$+\tau_{11}^{hj}$	$+\tau_{11}^{hk}$			0.1167
2121	τ_1^h		$+\tau_{11}^{hj}$		$+\tau_{11}^{h\ell}$		0.0390
2122	τ_1^h		$+\tau_{11}^{hj}$				-0.2290
2211	τ_1^h			$+\tau_{11}^{hk}$	$+\tau_{11}^{h\ell}$		0.2644
2212	τ_1^h			$+\tau_{11}^{hk}$			-0.0036
2221	τ_1^h				$+\tau_{11}^{h\ell}$		-0.0813
2222	τ_1^h						-0.3492

$\tau_1^h = -0.3492,\ \tau_{11}^{hi} = -0.4065,\ \tau_{11}^{hj} = 0.1203$

$\tau_{11}^{hk} = 0.3457,\ \tau_{11}^{h\ell} = 0.2680,\ \tau_{111}^{hi\ell} = 0.3789$

TABLE 3c

Log-odds $\ell n\ x^*_w(1ijk\ell)/x^*_w(2ijk\ell)$

$ijk\ell$	Parametric representation							log-odds
1111	τ^h_1	$+\tau^{hi}_{11}$	$+\tau^{hj}_{11}$	$+\tau^{hk}_{11}$	$+\tau^{h\ell}_{11}$	$+\tau^{hi\ell}_{111}$	$+\tau^{hj\ell}_{111}$	0.3283
1112	τ^h_1	$+\tau^{hi}_{11}$	$+\tau^{hj}_{11}$	$+\tau^{hk}_{11}$				-0.2523
1121	τ^h_1	$+\tau^{hi}_{11}$	$+\tau^{hj}_{11}$		$+\tau^{h\ell}_{11}$	$+\tau^{hi\ell}_{111}$	$+\tau^{hj\ell}_{111}$	-0.0197
1122	τ^h_1	$+\tau^{hi}_{11}$	$+\tau^{hj}_{11}$					-0.6004
1211	τ^h_1	$+\tau^{hi}_{11}$		$+\tau^{hk}_{11}$	$+\tau^{h\ell}_{11}$	$+\tau^{hi\ell}_{111}$		0.3976
1212	τ^h_1	$+\tau^{hi}_{11}$		$+\tau^{hk}_{11}$				0.5396
1221	τ^h_1	$+\tau^{hi}_{11}$			$+\tau^{h\ell}_{11}$	$+\tau^{hi\ell}_{111}$		0.0495
1222	τ^h_1	$+\tau^{hi}_{11}$						-0.8876
2111	τ^h_1		$+\tau^{hj}_{11}$	$+\tau^{hk}_{11}$	$+\tau^{h\ell}_{11}$		$+\tau^{hj\ell}_{111}$	0.3542
2112	τ^h_1		$+\tau^{hj}_{11}$	$+\tau^{hk}_{11}$				0.1512
2121	τ^h_1		$+\tau^{hj}_{11}$		$+\tau^{h\ell}_{11}$		$+\tau^{hj\ell}_{111}$	0.0061
2122	τ^h_1		$+\tau^{hj}_{11}$					-0.1968
2211	τ^h_1			$+\tau^{hk}_{11}$	$+\tau^{h\ell}_{11}$			0.4235
2212	τ^h_1			$+\tau^{hk}_{11}$				-0.1360
2221	τ^h_1				$+\tau^{h\ell}_{11}$			0.0754
2222	τ^h_1							-0.4841

$\tau^h_1 = -0.4841,\ \tau^{hi}_{11} = -0.4035,\ \tau^{hj}_{11} = 0.2873$

$\tau^{hk}_{11} = 0.3481,\ \tau^{h\ell}_{11} = 0.5595,\ \tau^{hi\ell}_{111} = 0.3776,\ \tau^{hj\ell}_{111} = -0.3565$

TABLE 3d

Odds-factors, Low IQ/High IQ, x_e^*

	Factors							
Base	Response 1		Response 2		Response 3		Response 4	
0.682	$i = 1$	0.816	$j = 1$	1.132	$k = 1$	1.406	$\ell = 1$	1.396
	$i = 2$	1.000	$j = 2$	1.000	$k = 2$	1.000	$\ell = 2$	1.000

The largest odds, low IQ/high IQ, occur for

$$x_e^*(12111)/x_e^*(22111) = (0.682)(1.000)(1.132)(1.406)(1.396)$$
$$= 1.515$$

The smallest odds, low IQ/high IQ, occur for

$$x_e^*(11222)/x_e^*(21222) = (0.682)(0.816)(1.000)(1.000)(1.000)$$
$$= 0.557$$

E_1: {ijkℓ: ℓn odds $\geqq$ 0}

TABLE 4a

E_1: Observations

ijkℓ	x(1ijkℓ)	x(2ijkℓ)
1111	82	53
1211	14	9
1221	11	10
2111	305	217
2112	200	172
2121	253	247
2211	41	25
2221	70	68
	976	801

$\mu_2(E_1) = 801/1491,$

$\mu_1(E_2) = (1491 - 976)/1491$

Prob. Error

$= (1/2)(801 + 515)/1491$

$= 1316/2982 = 0.441$

TABLE 4b

E_1: x_e^*

ijkℓ	x_e^*(1ijkℓ)	x_e^*(2ijkℓ)
1111	74.656	60.346
1211	12.010	10.990
2111	314.491	207.508
2112	193.625	178.376
2121	259.322	240.679
2211	37.780	28.219
	891.884	726.118

$\mu_2(E_1) = 726.118/1491,$

$\mu_1(E_2) = (1491 - 891.884)/1491$

Prob. Error

$= (1/2)(726.188 + 599.166)/1491$

$= 1325.234/2982 = 0.444$

TABLE 4c

E_1: x_v^*

ijkℓ	x_v^*(1ijkℓ)	x_v^*(2ijkℓ)
1111	79.426	55.574
1121	50.789	50.211
1211	12.855	10.144
2111	310.589	211.411
2112	196.841	175.160
2121	254.876	245.125
2211	37.337	28.662
	942.713	776.287

$\mu_2(E_1) = 776.287/1491$

$\mu_1(E_2) = (1491 - 942.713)/1491$

Prob. Error

$= (1/2)(776.287 + 548.287)/1491$

$= 1324.574/2982$

$= 0.444$

TABLE 4d

E_1: x_w^*

$ijk\ell$	$x_w^*(1ijk\ell)$	$x_w^*(2ijk\ell)$
1111	78.482	56.517
1211	13.756	9.244
1212	8.102	13.898
1221	10.760	10.240
2111	306.748	215.252
2112	200.037	171.963
2121	250.769	249.232
2211	39.884	26.115
2221	71.600	66.401
	980.138	818.862

$\mu_2(E_1) = 818.862/1491$

$\mu_1(E_2) = (1491 - 980.138)/1491$

Prob. Error

$= (1/2)(818.862 + 510.862)/1491$

$= 1329.724/2982$

$= 0.446$

TABLE 4e

Martin and Bradley

E_1	$\hat{x}(1ijk\ell)$	$\hat{x}(2ijk\ell)$
1111	74.67	60.33
1211	12.02	10.98
2111	314.50	207.50
2112	193.45	178.55
2121	259.17	240.83
2211	37.74	28.26
	891.55	726.45

$\mu_2(E_1) = 726.45/1491$,

$\mu_1(E_2) = (1491 - 891.55)/1491$

Prob. Error $= (1/2)(726.45 + 599.45)/1491$

$= 1325.90/2982$

$= 0.445$

TABLE 5

	1	2	3	4	5	6	7	8	9	10	11	12	13	14	15	16	17	18	19	20
hijkℓ	L	h	i	j	k	ℓ	hi	hj	hk	hℓ	ij	ik	iℓ	jk	jℓ	kℓ	hij	hik	hiℓ	hjk
		1	1	1	1	1	11	11	11	11	11	11	11	11	11	11	111	111	111	111
11111	1	1	1	1	1	1	1	1	1	1	1	1	1	1	1	1	1	1	1	1
11112	1	1	1	1	1		1	1	1		1	1		1			1	1		1
11121	1	1	1	1		1	1	1		1	1		1		1		1		1	
11122	1	1	1	1			1	1			1						1			
11211	1	1	1		1	1	1		1	1		1	1			1		1	1	
11212	1	1	1		1		1		1			1						1		
11221	1	1	1			1	1			1			1						1	
11222	1	1	1				1													
12111	1	1		1	1	1		1	1	1				1	1	1				1
12112	1	1		1	1			1	1					1						1
12121	1	1		1		1		1		1					1					
12122	1	1		1				1												
12211	1	1			1	1			1	1						1				
12212	1	1			1				1											
12221	1	1				1				1										
12222	1	1																		
21111	1		1	1	1	1					1	1	1	1	1	1				
21112	1		1	1	1						1	1		1						
21121	1		1	1		1					1		1		1					
21122	1		1	1							1									
21211	1		1		1	1						1	1			1				
21212	1		1		1							1								
21221	1		1			1							1							
21222	1		1																	
22111	1			1	1	1								1	1	1				
22112	1			1	1									1						
22121	1			1		1									1					
22122	1			1																
22211	1				1	1										1				
22212	1				1															
22221	1					1														
22222	1																			
x		√	√	√	√	√	√	√	√	√	√	√	√	√	√	√	√	√	√	√
x^*_e		√	√	√	√	√	√	√	√	√	√	√	√	√	√	√				
x^*_v		√	√	√	√	√	√	√	√	√	√	√	√	√	√	√			√	
x^*_w		√	√	√	√	√	√	√	√	√	√	√	√	√	√	√			√	

21	22	23	24	25	26	27	28	29	30	31	32
hjℓ	hkℓ	ijk	ijℓ	ikℓ	jkℓ	h i j k	h i j ℓ	h i k ℓ	h j k ℓ	i j k ℓ	h i j k ℓ
111	111	111	111	111	111						
1	1	1	1	1	1	1	1	1	1	1	1
		1				1					
1			1				1				
	1			1				1			
1	1	1			1				1		
1											
	1										
		1	1	1	1					1	
		1									
			1								
				1							
					1						
√	√	√	√	√	√	√	√	√	√	√	√
		√	√	√	√					√	
		√	√	√	√					√	
√		√	√	√	√					√	

TABLE 6

Covariance Matrix - Estimated Natural Parameters - Log-odds Representation x_e^*

	τ_1^h	τ_{11}^{hi}	τ_{11}^{hj}	τ_{11}^{hk}	$\tau_{11}^{h\ell}$
τ_1^h	0.0092	-0.0014	-0.0071	-0.0015	-0.0023
τ_{11}^{hi}		0.0095	0.0002	-0.0007	-0.0002
τ_{11}^{hj}			0.0092	-0.0008	-0.0003
τ_{11}^{hk}				0.0057	-0.0005
$\tau_{11}^{h\ell}$					0.0055

EXAMPLE 2

Leukemia death observations at Atomic Bomb Casualty Commission. This example illustrates the analysis of a three-way 5 x 6 x 2 contingency table and the MDI estimation procedure for the hypothesis of no second-order interaction. It also illustrates the use of a cell, other than the last one, as the reference cell. Details of the computation of the covariance matrix of a set of estimated natural parameters of interest are given, as well as the computation of confidence intervals using the multiple comparison lemma. We are grateful to Professor N. Sugiura for the data and permission to use it.

THE ANALYSIS OF LEUKEMIA DEATH OBSERVATIONS AT ATOMIC BOMB CASUALTY COMMISSION

Sugiura and Otake (1974) have considered the analysis of k 2 x c contingency tables and have applied their procedures to the data in Table 1. We propose to apply the MDI estimation and associated concepts to the analysis of the same data. We denote the occurrences in the three-way contingency Table 1 by x(ijk) with the notation

Characteristic	Index	1	2	3	4	5	6
Age	i	0-9	10-19	20-34	35-49	50+	
Dose	j	Not in city	0-9	10-49	50-99	100-199	200+
Mortality	k	Dead	Alive				

We get the MDI estimates fitting the sets of marginals

a) x(ij.), x(..k),

b) x(ij.), x(i.k),

c) x(ij.), x(i.k), x(.jk),

d) x(ij.), x(.jk).

We start with the set of marginals x(ij.), x(..k) because $x_a^*(ijk) = x(ij.)x(..k)/n$ is the MDI (or ML) estimate under the null hypothesis that mortality is homogeneous over the age by dose combinations. We summarize the results in the Analysis of Information Table.

TABLE

Analysis of Information

Component due to	Information	D.F.
a) x(ij.), x(..k)	$2I(x:x_a^*) = 205.983$	29
b) x(ij.), x(i.k)	$2I(x_b^*:x_a^*) = 2.326$	4
	$2I(x:x_b^*) = 203.657$	25

Analysis of Information (Continued)

	Component due to	Information	D.F.
c)	x(ij.), x(i.k), x(.jk)	$2I(x_c^*:x_b^*) = 175.810$	5
		$2I(x:x_c^*) = 27.847$	20
a)	x(ij.), x(..k)	$2I(x:x_a^*) = 205.983$	29
d)	x(ij.), x(.jk)	$2I(x_d^*:x_a^*) = 173.502$	5
		$2I(x:x_d^*) = 32.481$	24
c)	x(ij.), x(.jk), x(i.k)	$2I(x_c^*:x_d^*) = 4.634$	4
		$2I(x:x_c^*) = 27.847$	20

We make the following inferences from the Analysis of Information Table.

1. Mortality is not homogeneous over the age by dose combinations ($2I(x:x_a^*) = 205.983$, 29 D.F.).

2. The effects of age by mortality are not significant ($2I(x_b^*:x_a^*) = 2.326$, 4 D.F., $2I(x_c^*:x_d^*) = 4.634$, 4 D.F.).

3. The effects of dose by mortality are highly significant ($2I(x_c^*:x_b^*) = 175.810$, 5 D.F., $2I(x_d^*:x_a^*) = 173.502$, 5 D.F.).

Note the differences as a result of a different ordering (see the discussion in Chapter 3 Section VII A).

Since the value of $2I(x:x_d^*) = 32.481$, 24 D.F. is not significant at the 10% level, we obtained the MDI estimate x_d^* as shown in Table 2b. However since four OUTLIER values were indicated for x_d^*, and for comparison with the results of Sugiura and Otake, it was decided to perform a more complete analysis with the estimate fitting all the two-way marginals, that is, the MDI estimate corresponding to an hypothesis of no second-order interaction. This MDI estimate is given in Table 2a and we have called it $x_2^*(ijk)$, that is, $x_2^*(ijk) \equiv x_c^*(ijk)$.

Again for easier comparison with the results of Sugiura and Otake we selected the cell (512) as the reference cell so that the loglinear representation of $x_2^*(ijk)$ is given by

$$\begin{aligned} \ell n\ x_2^*(ijk)/n(1/60) = L &+ \tau_1^i T_1^i(ijk) + \ldots + \tau_4^i T_4^i(ijk) + \tau_2^j T_2^j(ijk) \\ &+ \ldots + \tau_6^j T_6^j(ijk) + \tau_1^k T_1^k(ijk) + \tau_{12}^{ij} T_{12}^{ij}(ijk) \\ &+ \ldots + \tau_{46}^{ij} T_{46}^{ij}(ijk) + \tau_{11}^{ik} T_{11}^{ik}(ijk) + \ldots + \\ &\tau_{41}^{ik} T_{41}^{ik}(ijk) + \tau_{21}^{jk} T_{21}^{jk}(ijk) + \ldots + \\ &\tau_{61}^{jk} T_{61}^{jk}(ijk) \end{aligned}$$

where the natural parameters (taus) are main effect and interaction parameters and the T(ijk) are the explanatory variables, the indicator functions of the corresponding marginals,

$$\text{e.g.} \sum_{ijk} T_{12}^{ij}(ijk) x_2^*(ijk) = x_2^*(12.) = x(12.) \text{ etc.}$$

From the loglinear representation of $x_2^*(ijk)$ we have the representation of the mortality log-odds or logit as

$$\ell n\ x_2^*(ij1)/x_2^*(ij2) = \tau_1^k + \tau_{i1}^{ik} + \tau_{j1}^{jk}$$

where $\tau_{51}^{ik} = 0 = \tau_{11}^{jk}$ by convention. We can evaluate the natural parameters, for example, as follows

$$\ell n\ x_2^*(511)/x_2^*(512) = \tau_1^k$$

$$\ell n\ x_2^*(111)/x_2^*(112) = \tau_1^k + \tau_{11}^{ik}$$

$$\cdots$$

$$\ell n\ x_2^*(411)/x_2^*(412) = \tau_1^k + \tau_{41}^{ik}$$

$$\ell n\ x_2^*(521)/x_2^*(522) = \tau_1^k + \tau_{21}^{jk}$$

$$\vdots$$

$$\ell n\ x_2^*(561)/x_2^*(562) = \tau_1^k + \tau_{61}^{jk}.$$

The following are the values obtained

$$\tau_1^k = -7.4714 \qquad \tau_{21}^{jk} = 0.5017$$

$$\tau_{11}^{ik} = -0.0849 \qquad \tau_{31}^{jk} = 0.9685$$

$$\tau_{21}^{ik} = -0.4515 \qquad \tau_{41}^{jk} = 1.2848$$

$$\tau_{31}^{ik} = -0.2655 \qquad \tau_{51}^{jk} = 2.2293$$

$$\tau_{41}^{ik} = 0.0371 \qquad \tau_{61}^{jk} = 3.4785$$

Sugiura and Otake (1974) used the representation for the log-odds

$$\log (p_{ij}/(1 - p_{ij})) = \mu + \alpha_i + \beta_j$$

where $\sum_{i=1}^{5} \alpha_i = 0$, $\beta_1 = 0$ and give the estimates

$$\hat{\alpha}_1 = 0.068 \qquad \hat{\beta}_2 = 0.502$$
$$\hat{\alpha}_2 = -0.299 \qquad \hat{\beta}_3 = 0.969$$
$$\hat{\alpha}_3 = -0.113 \qquad \hat{\beta}_4 = 1.285$$
$$\hat{\alpha}_4 = 0.190 \qquad \hat{\beta}_5 = 2.229$$
$$\hat{\alpha}_5 = 0.153 \qquad \hat{\beta}_6 = 3.478.$$

We note that $\tau_{21}^{jk} = \beta_2, \ldots, \tau_{61}^{jk} = \beta_6$ and

$$\mu + \alpha_5 = \tau_1^k \qquad \mu + \alpha_3 = \tau_1^k + \tau_{31}^{ik}$$
$$\mu + \alpha_1 = \tau_1^k + \tau_{11}^{ik} \qquad \mu + \alpha_4 = \tau_1^k + \tau_{41}^{ik},$$
$$\mu + \alpha_2 = \tau_1^k + \tau_{21}^{ik}$$

that is

$$\alpha_1 = \tau_{11}^{ik} - (\tau_{11}^{ik} + \tau_{21}^{ik} + \tau_{31}^{ik} + \tau_{41}^{ik})/5$$

$$\alpha_2 = \tau_{21}^{ik} - (\tau_{11}^{ik} + \tau_{21}^{ik} + \tau_{31}^{ik} + \tau_{41}^{ik})/5$$

$$\alpha_3 = \tau_{31}^{ik} - (\tau_{11}^{ik} + \tau_{21}^{ik} + \tau_{31}^{ik} + \tau_{41}^{ik})/5$$

$$\alpha_4 = \tau_{41}^{ik} - (\tau_{11}^{ik} + \tau_{21}^{ik} + \tau_{31}^{ik} + \tau_{41}^{ik})/5$$

$$\alpha_5 = -(\tau_{11}^{ik} + \tau_{21}^{ik} + \tau_{31}^{ik} + \tau_{41}^{ik})/5$$

yielding $\alpha_1 = 0.0680$, $\alpha_2 = -0.2986$, $\alpha_3 = -0.1126$, $\alpha_4 = 0.1900$, $\alpha_5 = 0.1529$.

We determine the covariance matrix of the natural parameters in the logit representation as follows. Let T denote the 60 x 40 matrix whose columns are

$$L,\ \overbrace{T_1^i(ijk),\ \ldots,\ T_4^i(ijk)}^{4},\ \overbrace{T_2^j(ijk),\ \ldots,\ T_6^j(ijk)}^{5},$$

$$\overbrace{T_{12}^{ij}(ijk),\ \ldots,\ T_{46}^{ij}(ijk)}^{20},\ T_1^k(ijk),$$

$$\underbrace{T_{11}^{ik}(ijk),\ \ldots,\ T_{41}^{ik}(ijk)}_{4},\ \underbrace{T_{21}^{jk}(ijk),\ \ldots,\ T_{61}^{jk}(ijk)}_{5}$$

and let $\underset{\sim}{D}$ denote a 60 x 60 diagonal matrix whose diagonal values are $x_2^*(ijk)$ in lexicographic order. Compute the 40 x 40 matrix $\underset{\sim}{S} = \underset{\sim}{T}'\underset{\sim}{D}\underset{\sim}{T}$

$$\underset{\sim}{S} = \begin{bmatrix} \underset{\sim}{S}_{11} & \underset{\sim}{S}_{12} \\ \underset{\sim}{S}_{21} & \underset{\sim}{S}_{22} \end{bmatrix}$$

where $\underset{\sim}{S}_{11}$ is 30 x 30 and $\underset{\sim}{S}_{22}$ is 10 x 10.

The covariance matrix of $\tau_1^k, \tau_{11}^{ik}, \ldots, \tau_{41}^{ik}, \tau_{21}^{jk}, \ldots, \tau_{61}^{jk}$ is

$$\underset{\sim}{S}_{22.1}^{-1} = (\underset{\sim}{S}_{22} - \underset{\sim}{S}_{21}\underset{\sim}{S}_{11}^{-1}\underset{\sim}{S}_{12})^{-1},$$

and is given in Table 3.

To compute confidence intervals for the natural parameters, following the procedure suggested by Sugiura and Otake using the multiple comparison lemma, Ferguson (1967, 282), we computed $(11.070 \text{ Var}(\tau_{j1}^{jk}))^{1/2}$ using the variances in Table 3 and obtained the following confidence intervals

τ_{21}^{jk}	-0.5463	1.5497
τ_{31}^{jk}	-0.2295	2.1665
τ_{41}^{jk}	-0.2762	2.8458
τ_{51}^{jk}	0.9233	3.5353
τ_{61}^{jk}	2.4185	4.5385.

The confidence intervals for the τ^{ik}'s were obtained by computing $(9.488 \text{ Var}(\tau_{i1}^{ik}))^{1/2}$ using the variances in Table 3 leading to

τ_{11}^{ik}	-0.9689	0.7991
τ_{21}^{ik}	-1.3485	0.4455

τ_{31}^{ik}	-1.1525	0.6215
τ_{41}^{ik}	-0.7759	0.8501

To relate with the bounds given by Sugiura and Otake for the α's, since we have seen that

$$\alpha_5 = -(\tau_{11}^{ik} + \tau_{21}^{ik} + \tau_{31}^{ik} + \tau_{41}^{ik})/5$$

we have that

$$\text{Var}(\alpha_5) = \frac{1}{25}\left(\text{Var}(\tau_{11}^{ik}) + \ldots + \text{Var}(\tau_{41}^{ik}) + 2\sum_{m<n} \text{cov}(\tau_{m1}^{ik}, \tau_{n1}^{ik})\right)$$

and from the entries in Table 3 we finally find $\text{Var}(\alpha_5) = 0.0339$, leading to the interval

$$\alpha_5 \quad -0.4141,\ 0.7199.$$

This agrees with the one obtained by Sugiura and Otake. Agreement with bounds given by Sugiura and Otake for other α's can be verified in a similar manner.

In the output corresponding to fitting all the two-way marginals, the entry corresponding to the cell (111) had a large OUTLIER value (5.239). Accordingly we computed an MDI estimate fitting all the two-way marginals but omitting the values x(111), x(112). This MDI estimate is denoted by $x_e^*(ijk)$ and its values are given in Table 2c.

The associated Analysis of Information is

Analysis of Information

Component due to	Information	D.F.
x(ij.), x(i.k), x(.jk)	$2I(x:x_2^*) = 27.847$	20
as above with $x_e^*(111) = x(111)$, $x_e^*(112) = x(112)$	$2I(x_e^*:x_2^*) = 6.223$	1
	$2I(x:x_e^*) = 21.614$	19

Removing x(111), x(112), that is x(11.) from the marginal estimation constraints gives an improved fit. We did not carry out any extensive analysis with $x_e^*(ijk)$ but did note the approximate equality of

$$\tau_{61}^{jk} - \tau_{51}^{jk}, \tau_{51}^{jk} - \tau_{41}^{jk}, \tau_{41}^{jk} - \tau_{31}^{jk}, \tau_{31}^{jk} - \tau_{21}^{jk}$$

when computed for x_e^* and x_2^*, the respective values being

x_e^*	x_2^*
1.249	1.249
0.949	0.944
0.320	0.316
0.466	0.467

TABLE 1

Original Data[1] x(ijk)

		Not in city		0-9		10-49	
		j = 1		j = 2		j = 3	
		dead	alive	dead	alive	dead	alive
Age	i	k = 1	k = 2	k = 1	k = 2	k = 1	k = 2
0-9	1	0	5015	7	10752	3	2989
10-19	2	5	5973	4	11811	6	2620
20-34	3	2	5669	8	10828	3	2798
35-49	4	3	6158	19	12645	4	3566
50+	5	3	3695	7	9053	3	2415
		13	26510	45	55089	19	14388

[1]From S. Jablon and H. Kato (1971), Mortality among A-bomb survivors, 1950-1970. Japanese National Institute of Health-Atomic Bomb Casualty Commission Life Span Study Report 6, ABCC TR 10-71.

TABLE 2a

Estimates

$x_2^*(ijk)$ Fitting marginals x(ij.), x(i.k), x(.jk)

	j = 1		j = 2		j = 3	
i	k = 1	k = 2	k = 1	k = 2	k = 1	k = 2
1	2.621	5012.379	9.282	10749.719	4.115	2987.887
2	2.165	5975.828	7.066	11807.930	2.504	2623.496
3	2.474	5668.520	7.804	10828.191	3.216	2797.783
4	3.637	6157.359	12.341	12651.648	5.546	3564.455
5	2.103	3695.898	8.507	9051.488	3.619	2414.382

50-99		100-199		200+	
j = 4		j = 5		j = 6	
dead	alive	dead	alive	dead	alive
k = 1	k = 2	k = 1	k = 2	k = 1	k = 2
1	694	4	418	11	387
1	771	3	792	6	820
1	797	3	596	9	624
2	972	1	694	10	608
2	655	2	393	6	289
7	3889	13	2893	42	2728

j = 4		j = 5		j = 6	
k = 1	k = 2	k = 1	k = 2	k = 1	k = 2
1.311	693.689	2.040	419.960	6.632	391.369
1.010	770.990	2.668	792.332	9.588	816.411
1.257	796.743	2.420	596.580	8.829	624.170
2.075	971.925	3.794	691.206	11.608	606.391
1.349	655.652	2.078	392.922	5.343	289.657

TABLE 2b

$x_d^*(ijk)$ Fitting marginals x(ij.), x(.jk),

$x_d^*(ijk) = x(ij.)x(.jk)/x(.j.)$

	j = 1		j = 2		j = 3	
i	k = 1	k = 2	k = 1	k = 2	k = 1	k = 2
1	2.458	5012.543	8.781	10750.215	3.946	2988.055
2	2.930	5975.070	9.643	11805.352	3.463	2622.537
3	2.780	5668.219	8.844	10827.152	3.694	2797.307
4	3.020	6157.980	10.336	12653.660	4.708	3565.293
5	1.813	3696.189	7.395	9052.602	3.189	2414.812

TABLE 2c

$x_e^*(ijk)$ - one outlier removed from $x_2^*(ijk)$,

that is x(111), x(112) or x(11.)

	j = 1		j = 2		j = 3	
i	k = 1	k = 2	k = 1	k = 2	k = 1	k = 2
1	0	5015	10.303	10748.695	4.561	2987.441
2	2.715	5975.285	6.870	11808.125	2.431	2623.570
3	3.095	5667.902	7.573	10828.422	3.117	2797.884
4	4.554	6156.441	11.984	12652.012	5.378	3564.625
5	2.636	3695.365	8.270	9051.727	3.514	2414.487

j = 4		j = 5		j = 6	
k = 1	k = 2	k = 1	k = 2	k = 1	k = 2
1.249	693.751	1.888	420.112	6.035	391.965
1.387	770.613	3.556	791.443	12.524	813.476
1.434	796.566	2.679	596.320	9.598	623.402
1.750	972.250	3.109	691.891	9.370	608.629
1.180	655.819	1.767	393.233	4.473	290.527

j = 4		j = 5		j = 6	
k = 1	k = 2	k = 1	k = 2	k = 1	k = 2
1.459	693.541	2.279	419.720	7.398	390.602
0.984	771.015	2.612	792.388	9.388	816.612
1.223	796.777	2.364	596.635	8.628	624.371
2.020	971.980	3.710	691.290	11.354	606.645
1.314	655.685	2.034	392.966	5.231	289.769

TABLE 3

Covariance Matrix τ_1^k, τ_{11}^{ik}, τ_{21}^{ik}, τ_{31}^{ik}, τ_{41}^{ik}, τ_{21}^{jk}, τ_{31}^{jk}, τ_{41}^{jk}, τ_{51}^{jk}, τ_{61}^{jk} of x_2^*

	τ_1^k	τ_{11}^{ik}	τ_{21}^{ik}	τ_{31}^{ik}	τ_{41}^{ik}
τ_1^k	0.1140	-0.0445	-0.0438	-0.0443	-0.0441
τ_{11}^{ik}		0.0824	0.0438	0.0438	0.0438
τ_{21}^{ik}			0.0849	0.0444	0.0441
τ_{31}^{ik}				0.0829	0.0440
τ_{41}^{ik}					0.0697
τ_{21}^{jk}					
τ_{31}^{jk}					
τ_{41}^{jk}					
τ_{51}^{jk}					
τ_{61}^{jk}					

τ_{21}^{jk}	τ_{31}^{jk}	τ_{41}^{jk}	τ_{51}^{jk}	τ_{61}^{jk}
-0.0782	-0.0782	-0.0783	-0.0769	-0.0755
0.0010	0.0007	0.0019	0.0016	0.0002
0.0016	0.0027	0.0023	-0.0017	-0.0041
0.0019	0.0021	0.0018	0.0000	-0.0024
0.0013	0.0010	0.0009	-0.0004	-0.0014
0.0993	0.0770	0.0770	0.0769	0.0769
	0.1298	0.0770	0.0769	0.0768
		0.2202	0.0770	0.0769
			0.1544	0.0771
				0.1015

EXAMPLE 3

Automobile accident data. This example illustrates the analysis of a four-way 3 x 4 x 3 x 2 contingency table. It points out that the model fitted determines the form of the log-odds or logit representation, but the converse is not true. The covariance matrix of the estimated natural parameters (taus) is given.

AUTOMOBILE ACCIDENT DATA - DRIVER EJECTION

Data used in this example are taken from a study of the relationship between car size and accident injuries as given in Kihlberg, Narragon and Campbell (1964). The observed data are given in Table 1 and the observed occurrences are denoted by $x(ijk\ell)$ where

Characteristic	Index	1	2	3	4
Car weight	i	Small	Compact	Standard	
Accident type	j	Collision with vehicle	Collision with object	Rollover without collision	Other roll-over
Severity	k	Not severe	Mod. severe	Severe	
Driver Ejection	ℓ	Not ejected	Ejected		

A condensed 2 x 2 x 2 x 2 version of this data was studied by Bhapkar and Koch (1968) and Ku and Kullback (1968).

Since the question of interest is the possible relation of driver ejection on car weight, accident type and severity, we start the fitting sequence with the marginals $x(ijk.)$, $x(...\ell)$. This first estimate, $x_a^*(ijk\ell) = x(ijk.)x(...\ell)/n$, corresponds to a null hypothesis that driver ejection is homogeneous over the 36 combinations of the other characteristics. As may be seen from the analysis of information table this hypothesis is clearly rejected by the data. It is found that fitting the model incorporating in addition to $x(ijk.)$ the marginals $x(i..\ell)$, $x(.j.\ell)$, $x(..k\ell)$, that is, the interactions of car weight, accident type, and severity respectively with driver ejection, a satisfactory fit to the observed data is obtained. The models fitting in addition three-way marginals $x(ij.\ell)$, etc., showed no significant effects for the associated natural parameters. The results are summarized in the analysis of information table.

TABLE

Analysis of Information

Component due to	Information	D.F.
a) $x(ijk.)$, $x(\ldots\ell)$	$2I(x:x_a^*) = 613.102$	35
b) $x(ijk.)$, $x(i..\ell)$, $x(.j.\ell)$,	$2I(x_b^*:x_a^*) = 587.584$	7
$x(..k\ell)$	$2I(x:x_b^*) = 25.518$	28
c) $x(ijk.)$, $x(ij.\ell)$, $x(i.k\ell)$,	$2I(x_c^*:x_b^*) = 14.491$	16
$x(.jk\ell)$	$2I(x:x_c^*) = 11.028$	12

The fitted MDI values $x_b^*(ijk\ell)$ are given in Table 2. The loglinear representation of $x_b^*(ijk\ell)$ contains the natural parameters L (a normalizing constant), τ_1^i, τ_2^i, τ_1^j, τ_2^j, τ_3^j, τ_1^k, τ_2^k, τ_1^ℓ, τ_{11}^{ij}, τ_{12}^{ij}, τ_{13}^{ij}, τ_{21}^{ij}, τ_{22}^{ij}, τ_{23}^{ij}, τ_{11}^{ik}, τ_{12}^{ik}, τ_{21}^{ik}, τ_{22}^{ik}, $\tau_{11}^{i\ell}$, $\tau_{21}^{i\ell}$, τ_{11}^{jk}, τ_{12}^{jk}, τ_{21}^{jk}, τ_{22}^{jk}, τ_{31}^{jk}, τ_{32}^{jk}, $\tau_{11}^{j\ell}$, $\tau_{21}^{j\ell}$, $\tau_{31}^{j\ell}$, $\tau_{11}^{k\ell}$, $\tau_{21}^{k\ell}$, τ_{111}^{ijk}, τ_{112}^{ijk}, τ_{121}^{ijk}, τ_{122}^{ijk}, τ_{131}^{ijk}, τ_{132}^{ijk}, τ_{211}^{ijk}, τ_{212}^{ijk}, τ_{221}^{ijk}, τ_{222}^{ijk}, τ_{231}^{ijk}, τ_{232}^{ijk}. The 28 additional natural parameters which would appear in the complete model for $x(ijk\ell)$ are hypothesized as zero and represent the 28 degrees of freedom of $2I(x:x_b^*)$. The log-odds or logit representation for the MDI estimate x_b^* is

$$\ell n\ x_b^*(ijk1)/x_b^*(ijk2) = \tau_1^\ell + \tau_{i1}^{i\ell} + \tau_{j1}^{j\ell} + \tau_{k1}^{k\ell}.$$

Parameters not involving ℓ are common to the loglinear representation of the MDI estimates of the cells of the odds and drop out. The values of the natural parameters may be obtained as

$$\tau_1^{\ell} = \ell n\ x_b^*(3431)/x_b^*(3432)$$

$$\tau_{11}^{i\ell} = \ell n\ x_b^*(1431)/x_b^*(1432) - \tau_1^{\ell}$$

$$\tau_{21}^{i\ell} = \ell n\ x_b^*(2431)/x_b^*(2432) - \tau_1^{\ell}$$

etc.

and

$\tau_1^{\ell} = -0.0083$ $\tau_{11}^{j\ell} = 1.3665$ $\tau_{11}^{k\ell} = 1.6085$

$\tau_{11}^{i\ell} = -0.2936$ $\tau_{21}^{j\ell} = 1.1139$ $\tau_{21}^{k\ell} = 0.8823$

$\tau_{21}^{i\ell} = -0.0788$ $\tau_{31}^{j\ell} = -0.2405.$

We recall that any natural parameter with a subscript $i = 3$ and/or $j = 4$ and/or $k = 3$ and/or $\ell = 2$ is by convention zero.

It is important to note that the MDI estimate $x_2^*(ijk\ell)$ obtained by fitting the two-way marginals $x(ij..)$, $x(i.k.)$, $x(i..\ell)$, $x(.jk.)$, $x(.j.\ell)$, $x(..k\ell)$ is one of several other MDI estimates that also have the log-odds or logit representation of the form

$$\ell n\ x_2^*(ijk1)/x_2^*(ijk2) = \tau_1^{\ell} + \tau_{i1}^{i\ell} + \tau_{j1}^{j\ell} + \tau_{k1}^{k\ell}.$$

The values of the natural parameters above depend however on the values of the MDI estimate $x_2^*(ijk\ell)$. The MDI estimate x_b^* makes no assumptions about the relationship among the explanatory variables or their associated natural parameters, and is to be preferred.

The model fitted determines the form of the log-odds or logit representation but the converse is not true.

For easier interpretation of the numerical values we use the representation of the estimated odds as the multiplicative model

$$x_b^*(ijk1)/x_b^*(ijk2) = \exp(\tau_1^{\ell})\ \exp(\tau_{i1}^{i\ell})\ \exp(\tau_{j1}^{j\ell})\ \exp(\tau_{k1}^{k\ell}).$$

The factors which determine the odds of Not ejected for any combination of the characteristics are:

FACTORS

Base	Car weight		Accident type		Severity	
0.99	Small	0.75	Collision with vehicle	3.92	Not severe	5.00
	Compact	0.92	Collision with object	3.05	Mod. severe	2.42
	Standard	1.00	Rollover without collision	0.79	Severe	1.00
			Other rollover	1.00		

By selecting the combination of characteristics with the largest factors, it is seen that the best odds for Not ejected, 19.40, occur for

Standard, Collision with vehicle, Not severe.

By selecting the combination of characteristics with the smallest factors, it is seen that the worst odds for Not ejected, 0.59, occur for

Small, Rollover without collision, Severe.

The observed odds for Not ejected from the original data are 4124/707 = 5.83. The estimated odds for any combination of characteristics is easily obtained from the factors above or the values of x_b^*.

The covariance matrix of the natural parameters in the logit representation for the MDI estimate x_b^* is given in Table 3. The computation is similar to that given in Example 2.

TABLE 1

Accident Data - Drivers Alone - Observed

Accident type	Accident severity	Not Ejected			Ejected		
		Small	Compact	Standard	Small	Compact	Standard
Collision with vehicle	Not severe	95	166	1279	8	7	65
	Mod. severe	31	34	506	2	5	51
	Severe	11	17	186	4	5	54
Collision with object	Not severe	34	55	599	5	6	46
	Mod. severe	8	34	241	2	4	26
	Severe	5	10	89	0	1	30
Rollover without Collision	Not severe	23	18	65	6	5	11
	Mod. severe	22	17	118	18	9	68
	Severe	5	2	23	5	6	33
Other Rollover	Not severe	9	10	83	6	2	11
	Mod. severe	23	26	177	13	16	78
	Severe	8	9	86	7	6	86
		274	398	3452	76	72	559

From Kihlberg, Narragon, and Campbell (1964).

TABLE 2

Accident data - Drivers Alone - MDI Estimate x_b^*

Accident type	Accident severity	Not ejected			Ejected		
		Small	Compact	Standard	Small	Compact	Standard
Collision with vehicle	Not severe	96.349	163.874	1278.209	6.651	9.126	65.790
	Mod. severe	28.879	34.973	503.433	4.121	4.027	53.567
	Severe	11.154	17.212	190.913	3.846	4.788	49.087
Collision with object	Not severe	35.817	56.919	604.917	3.183	4.081	40.082
	Mod. severe	8.448	33.095	234.832	1.552	4.905	32.167
	Severe	3.463	8.099	89.406	1.537	2.901	29.594
Rollover without Collision	Not severe	21.572	18.000	60.475	7.428	5.000	15.525
	Mod. severe	23.367	16.516	121.512	16.633	9.484	64.488
	Severe	3.676	3.351	24.535	6.324	4.649	31.465
Other Rollover	Not severe	11.804	9.849	78.213	3.196	2.151	15.787
	Mod. severe	23.082	28.936	179.924	12.918	13.064	75.076
	Severe	6.377	7.174	85.645	8.623	7.826	86.355

TABLE 3

Covariance matrix - natural parameters in the logit representation for the MDI estimate x_b^*

τ_1^{ℓ}	$\tau_{11}^{i\ell}$	$\tau_{21}^{i\ell}$	$\tau_{11}^{j\ell}$	$\tau_{21}^{j\ell}$	$\tau_{31}^{j\ell}$	$\tau_{11}^{k\ell}$	$\tau_{21}^{k\ell}$
.0017	.0003	.0003	.0005	.0003	.0003	.0005	.0003
	.0039	-.0003	.0000	-.0001	.0005	.0001	.0001
		.0027	.0001	.0000	.0001	.0001	.0000
			.0008	-.0005	-.0004	.0003	.0000
				.0012	-.0003	.0002	.0000
					.0036	-.0001	.0003
						.0008	-.0006
							.0011

EXAMPLE 4

Minnesota high-school graduates of June 1938. This example illustrates the analysis of a four-way 2 x 3 x 7 x 4 contingency table. In particular the dependent classification is not dichotomous as in the previous examples but has four categories. The final model leads to log-odds representations involving main effects and interactions.

CLASSIFICATION OF MINNESOTA HIGH-SCHOOL GRADUATES OF JUNE 1938

The data of this 2 x 3 x 7 x 4 contingency table represent a four-way cross classification of the April 1939 status of 13,968 Minnesota High-School graduates of June 1938. The data were presented by Hoyt, Krishnaiah, and Torrance (1959). They formulated and tested various hypotheses of independence using chi-square statistics. The same data were also used by Kullback, Kupperman, and Ku (1962b) to illustrate the use of the MDI statistics in the analysis of various hypotheses of independence and homogeneity. Patil (1974) condensed the original data into a 4 x 3 x 7 table by summing over the sex classification and tested for no second-order interaction in the three-way table by a global chi-square statistic.

We shall examine models fitting certain sets of marginals and analyze the data on the basis of the loglinear representation of a model that is a good fit to the data. The original data are listed in Table 1 where we denote the occurrences in the cells by x(hijk), with

Characteristic	Index	1	2	3	4	5	6	7
Sex	h	Male	Female					
H.S. Rank	i	Lowest third	Middle third	Upper third				
Father's Occupational Level	j	1	2	3	4	5	6	7
Post H.S. Status	k	Enrolled in College	Noncollegiate school	Employed full time	Other			

The problem is to determine the relationship of post high-school status on the other variables. Note that here the dependent variable is polychotomous. We summarize in the analysis of information Table 4, the results of fitting three models to the data, or the sets of marginals,

$$H_a\text{: } x(hij.),\ x(...k),$$

$$H_b\text{: } x(hij.),\ x(h..k),\ x(.i.k),\ x(..jk),$$

$$H_c\text{: } x(hij.),\ x(.i.k),\ x(h.jk).$$

The MDI estimate x_a^*, corresponding to H_a, is to determine whether the occurrences of post high-school status are homogeneously distributed over the 42 combinations of sex, high-school rank, and father's occupational level. We note that $x_a^*(hijk) = x(hij.)x(...k)/n$. Since the data do not support the null hypothesis of homogeneity we consider the MDI estimate x_b^* corresponding to H_b. This estimate will provide a log-odds or logit representation in terms of a linear combination of the main effects of sex, high-school rank and father's occupational level on post high-school status. Since the fit of the MDI estimate x_b^* to the data was not considered satisfactory the effects of various natural parameters associated with three-way marginals were examined. The natural parameters with the largest effect, for the additional degrees of freedom, turned out to be that of sex x father's occupational level x post high-school status, that is, associated with the marginal x(h.jk). It was decided to analyze the data in terms of the MDI estimate x_c^* corresponding to H_c. The values of $x_c^*(hijk)$ are listed in Table 2.

From the loglinear representation of the MDI estimate x_c^*, we arrive at the following multiple representation for the log-odds

$$\ell n\ x_c^*(hij1)/x_c^*(hij4) = \tau_1^k + \tau_{h1}^{hk} + \tau_{i1}^{ik} + \tau_{j1}^{jk} + \tau_{hj1}^{hjk},$$

$$\ell n\ x_c^*(hij2)/x_c^*(hij4) = \tau_2^k + \tau_{h2}^{hk} + \tau_{i2}^{ik} + \tau_{j2}^{jk} + \tau_{hj2}^{hjk},$$

$$\ell n\ x_c^*(hij3)/x_c^*(hij4) = \tau_3^k + \tau_{h3}^{hk} + \tau_{i3}^{ik} + \tau_{j3}^{jk} + \tau_{hj3}^{hjk}.$$

The values of the natural parameters in the log-odds representations are:

$\tau_1^k = -1.0345$ $\qquad$ $\tau_2^k = -2.2548$ $\qquad$ $\tau_3^k = -1.7189$

$\tau_{11}^{hk} = 0.9935$ $\qquad$ $\tau_{12}^{hk} = -0.3523$ $\qquad$ $\tau_{13}^{hk} = -0.1111$

$\tau_{11}^{ik} = -1.5908$ $\qquad$ $\tau_{12}^{ik} = -1.0060$ $\qquad$ $\tau_{13}^{ik} = -1.0682$

$\tau_{21}^{ik} = -0.8912$ $\qquad$ $\tau_{22}^{ik} = -0.4542$ $\qquad$ $\tau_{23}^{ik} = -0.4934$

$\tau_{11}^{jk} = 2.2731$ $\qquad$ $\tau_{12}^{jk} = 0.9905$ $\qquad$ $\tau_{13}^{jk} = 0.8593$

$\tau_{21}^{jk} = 1.2332$ $\tau_{22}^{jk} = 0.9822$ $\tau_{23}^{jk} = 0.6872$

$\tau_{31}^{jk} = 0.4009$ $\tau_{32}^{jk} = 0.3932$ $\tau_{33}^{jk} = 0.6333$

$\tau_{41}^{jk} = 1.1259$ $\tau_{42}^{jk} = 0.8881$ $\tau_{43}^{jk} = 0.6099$

$\tau_{51}^{jk} = 0.6194$ $\tau_{52}^{jk} = 0.3995$ $\tau_{53}^{jk} = 0.5254$

$\tau_{61}^{jk} = -0.0321$ $\tau_{62}^{jk} = -0.1397$ $\tau_{63}^{jk} = 0.1989$

$\tau_{111}^{hjk} = -0.7277$ $\tau_{112}^{hjk} = -1.3054$ $\tau_{113}^{hjk} = -0.4037$

$\tau_{121}^{hjk} = -0.6340$ $\tau_{122}^{hjk} = -0.8018$ $\tau_{123}^{hjk} = -0.3643$

$\tau_{131}^{hjk} = -1.0923$ $\tau_{132}^{hjk} = -0.8030$ $\tau_{133}^{hjk} = -0.9709$

$\tau_{141}^{hjk} = -0.8463$ $\tau_{142}^{hjk} = -0.7581$ $\tau_{143}^{hjk} = -0.5573$

$\tau_{151}^{hjk} = -0.6402$ $\tau_{152}^{hjk} = -0.8605$ $\tau_{153}^{hjk} = -0.5503$

$\tau_{161}^{hjk} = -0.7537$ $\tau_{162}^{hjk} = -0.2334$ $\tau_{163}^{hjk} = -0.4397$

All natural parameters with subscripts $h = 2$ and/or $i = 3$ and/or $j = 7$ and/or $k = 4$ are zero by convention.

From the representations for the log-odds it is seen that the association between high-school rank and post high-school status is the same for all combinations of sex and father's occupational level, that is,

$$\ell n\ x_c^*(h1j1)/x_c^*(h1j4) - \ell n\ x_c^*(h2j1)/x_c^*(h2j4)$$

$$= \ell n\ x_c^*(h1j1)x_c^*(h2j4)/x_c^*(h1j4)x_c^*(h2j1)$$

$$= \tau_{11}^{ik} - \tau_{21}^{ik} = -0.6996,$$

$$\ell n\, x_c^*(h2j1)x_c^*(h3j4)/x_c^*(h2j4)x_c^*(h3j1) = \tau_{21}^{ik} = -0.8912,$$

$$\ell n\, x_c^*(h1j2)x_c^*(h2j4)/x_c^*(h1j4)x_c^*(h2j2) = \tau_{12}^{ik} - \tau_{22}^{ik} = -0.5518,$$

$$\ell n\, x_c^*(h2j2)x_c^*(h3j4)/x_c^*(h2j4)x_c^*(h3j2) = \tau_{22}^{ik} = -0.4542,$$

$$\ell n\, x_c^*(h1j3)x_c^*(h2j4)/x_c^*(h1j4)x_c^*(h2j3) = \tau_{13}^{ik} - \tau_{23}^{ik} = -0.5748,$$

$$\ell n\, x_c^*(h2j3)x_c^*(h3j4)/x_c^*(h2j4)x_c^*(h3j3) = \tau_{23}^{ik} = -0.4934.$$

The association between sex and post high-school status varies with father's occupational level, that is,

$$\ell n\, x_c^*(1ij1)/x_c^*(1ij4) - \ell n\, x_c^*(2ij1)/x_c^*(2ij4) = \tau_{11}^{hk} + \tau_{1j1}^{hjk},$$

$$\ell n\, x_c^*(1ij2)/x_c^*(1ij4) - \ell n\, x_c^*(2ij2)/x_c^*(2ij4) = \tau_{12}^{hk} + \tau_{1j2}^{hjk},$$

$$\ell n\, x_c^*(1ij3)/x_c^*(1ij4) - \ell n\, x_c^*(2ij3)/x_c^*(2ij4) = \tau_{13}^{hk} + \tau_{1j3}^{hjk}.$$

We summarize the numerical values below.

j	$\tau_{11}^{hk} + \tau_{1j1}^{hjk}$	$\tau_{12}^{hk} + \tau_{1j2}^{hjk}$	$\tau_{13}^{hk} + \tau_{1j3}^{hjk}$
1	0.2658	-1.6577	-0.5148
2	0.3595	-1.1541	-0.4754
3	-0.0988	-1.1603	-1.0820
4	0.1472	-1.1104	-0.6684
5	0.3533	-1.2128	-0.6619
6	0.2348	-0.5857	-0.5508
7	0.9935	-0.3523	-0.1111

We remark that father's occupational level 3 shows a peculiarity as compared to other values in the first column above. Kullback, Kupperman, and Ku (1962b, 593) noted that there was an unusually larger number of girls than boys for the third category of father's occupation. Apparently there was a tendency for the girls not to enroll in college as compared to the boys. In particular, for example, the association between sex and collegiate or noncollegiate school is

$$\ell n\ x_c^*(1ij1)/x_c^*(1ij2) - \ell n\ x_c^*(2ij1)/x_c^*(2ij2)$$

$$= \tau_{11}^{hk} + \tau_{1j1}^{hjk} - \tau_{12}^{hk} - \tau_{1j2}^{hjk}.$$

From the preceding results we have

j	$\tau_{11}^{hk} + \tau_{1j1}^{hjk} - \tau_{12}^{hk} - \tau_{1j2}^{hjk}$
1	1.9235
2	1.5136
3	1.0615
4	1.2576
5	1.5661
6	0.8205
7	1.3458

The association between father's occupational level and post high-school status varies with sex, that is,

$$\ell n\ x_c^*(hi11)/x_c^*(hi14) - \ell n\ x_c^*(hi71)/x_c^*(hi74) = \tau_{11}^{jk} + \tau_{h11}^{hjk},$$

$$\ell n\ x_c^*(hi21)/x_c^*(hi24) - \ell n\ x_c^*(hi71)/x_c^*(hi74) = \tau_{21}^{jk} + \tau_{h21}^{hjk},$$

etc.

$$\ell n\ x_c^*(hi12)/x_c^*(hi14) - \ell n\ x_c^*(hi72)/x_c^*(hi74) = \tau_{12}^{jk} + \tau_{h12}^{hjk},$$

$$\ell n\ x_c^*(hi22)/x_c^*(hi24) - \ell n\ x_c^*(hi72)/x_c^*(hi74) = \tau_{22}^{jk} + \tau_{h22}^{hjk},$$

etc.

$$\ell n\ x_c^*(hi13)/x_c^*(hi14) - \ell n\ x_c^*(hi73)/x_c^*(hi74) = \tau_{13}^{jk} + \tau_{h13}^{hjk},$$

$$\ell n\ x_c^*(hi23)/x_c^*(hi24) - \ell n\ x_c^*(hi73)/x_c^*(hi74) = \tau_{23}^{jk} + \tau_{h23}^{hjk},$$

etc.

A tabulation of these associations is

	h = 1			h = 2		
j	k = 1	k = 2	k = 3	k = 1	k = 2	k = 3
1	1.5094	-0.3149	0.4556	2.2731	0.9905	0.8593
2	0.5992	0.1804	0.3229	1.2332	0.9822	0.6872
3	-0.6914	-0.4148	-0.3376	0.4009	0.3932	0.6333
4	0.2796	0.1300	0.0526	1.1259	0.8881	0.6099
5	-0.0208	-0.4610	-0.0254	0.6194	0.3995	0.5254
6	-0.7908	-0.3731	-0.2408	-0.0321	-0.1397	0.1989

In particular, the association between father's occupational levels 1 and 2 and post high-school status of collegiate and noncollegiate school, for boys, is

$$\ln x_c^*(1i11)/x_c^*(1i12) - \ln x_c^*(1i21)/x_c^*(1i22) = \tau_{11}^{jk}$$

$$+ \tau_{111}^{hjk} - \tau_{12}^{jk} - \tau_{112}^{hjk} - \tau_{21}^{jk} - \tau_{121}^{hjk} + \tau_{22}^{jk} + \tau_{122}^{hjk}.$$

We shall not pursue this matter any further here. The reader should be able to examine any particular associations of interest.

We note from Analysis of Information Table 4 that $2I(x_b^*:x_a^*)/2I(x:x_a^*) = 2672.724/2824.434 = 0.946$, that is, the MDI estimate x_b^* leaves only 5.4% of the total variation $2I(x:x_a^*)$ unexplained, even though the fit as measured by the chi-square distribution was considered poor. This phenomenon occurs when the total number of observations is large, in this case 13,968. The MDI estimate x_c^* leaves only 3.5% of the total variation unexplained.

In view of this, it was deemed useful to list the values of the MDI estimate x_b^* in Table 3. The multiple parametric representations for the log-odds in terms of x_b^* and the associated values of the natural parameters are given on page 131.

We leave it as an exercise for the reader to examine and compare the effect of the interactions on the associations as given by x_c^* and the associations as given by x_b^*.

TABLE 1

Frequency for each Sex x High-School Rank x Father's Occupational Level x Post High-School Status Combination

x(hijk)

			High-School Rank											
			Lowest Third				Middle Third				Upper Third			
Post High-School Status*			1	2	3	4	1	2	3	4	1	2	3	4
Sex (1) Male	Father's Occupational Levels	1	87	3	17	105	216	4	14	118	256	2	10	53
		2	72	6	18	209	159	14	28	227	176	8	22	95
		3	52	17	14	541	119	13	44	578	119	10	33	257
		4	88	9	14	328	158	15	36	304	144	12	20	115
		5	32	1	12	124	43	5	7	119	42	2	7	56
		6	14	2	5	148	24	6	15	131	24	2	4	61
		7	20	3	4	109	41	5	13	88	32	2	4	41
Sex (2) Female	Father's Occupational Levels	1	53	7	13	76	163	30	28	118	309	17	38	89
		2	36	16	11	111	116	41	53	214	225	49	68	210
		3	52	28	49	521	162	64	129	708	243	79	184	448
		4	48	18	29	191	130	47	62	305	237	57	63	219
		5	12	5	10	101	35	11	37	152	72	20	21	95
		6	9	1	15	130	19	13	22	174	42	10	19	105
		7	3	1	6	88	25	9	15	158	36	14	19	93

*Categories of post high-school status: (1) enrolled in college; (2) enrolled in non-collegiate school; (3) employed full-time; (4) other.

From Hoyt, Krishnaiah, and Torrance (1959)

TABLE 2

Estimated Frequency for each Sex x High-School Rank x Father's Occupational Level x Post High-School Status Combination

			Lowest Third				
Post High-School Status*			1	2	3	4	1
Sex (1) Male	Father's Occupational Levels	1	96.076	2.062	9.106	104.751	214.142
		2	74.160	6.726	15.853	208.256	160.787
		3	52.918	9.622	21.244	540.216	114.549
		4	84.275	10.004	18.926	325.787	159.270
		5	25.645	2.277	7.194	133.882	44.415
		6	13.027	2.727	6.363	146.881	24.402
		7	20.818	2.871	5.867	106.443	37.619
Sex (2) Female	Father's Occupational Levels	1	53.675	7.884	11.104	76.337	168.174
		2	29.353	12.096	14.462	118.093	111.908
		3	54.868	28.839	58.878	507.420	151.976
		4	44.660	18.647	22.674	200.023	134.884
		5	13.289	5.649	10.289	98.774	40.924
		6	9.223	4.386	9.883	131.508	24.155
		7	6.054	3.206	5.149	83.592	22.830

*Categories of post high-school status: (1) enrolled in college (2) enrolled in non-collegiate school; (3) employed full-time; (4) other

$x_c^*(hijk)$

High-School Rank

Middle Third			Upper Third			
2	3	4	1	2	3	4
3.964	17.913	115.981	248.782	2.975	13.981	55.269
12.579	30.337	224.296	172.053	8.695	21.809	98.448
17.964	40.588	580.899	122.534	12.414	29.169	254.885
16.308	31.570	305.852	146.455	9.687	19.503	115.361
3.401	10.997	115.186	46.939	2.322	7.808	49.932
4.406	10.520	136.671	24.571	2.866	7.117	56.447
4.474	9.357	95.549	34.562	2.655	5.776	36.008
21.306	30.708	118.813	303.151	24.810	37.188	87.850
39.776	48.665	223.655	235.739	54.129	68.873	193.253
68.898	143.943	698.185	250.157	73.263	159.179	471.394
48.576	60.442	300.100	235.456	54.778	70.884	214.878
15.005	27.967	151.105	64.787	15.346	29.744	98.121
9.909	22.846	171.091	36.622	9.705	23.271	106.401
10.430	17.139	156.602	35.117	10.364	17.712	98.806

TABLE 3

Estimated Frequency for each Sex x High-School Rank x Father's Occupational Level x Post High-School Status Combination

		Lowest Third				
Post High-School Status*		1	2	3	4	1
Sex (1) Male	Father's Occupational Levels 1	94.781	3.276	8.127	105.816	211.498
	2	70.740	7.238	14.056	212.964	154.477
	3	61.853	9.484	27.106	525.548	130.626
	4	86.886	9.856	18.264	323.990	162.857
	5	24.134	2.512	7.083	135.289	41.884
	6	13.033	1.857	5.776	148.331	24.407
	7	15.184	1.813	4.304	114.698	28.631
Sex (2) Female	Father's Occupational Levels 1	54.832	7.145	11.839	75.181	170.390
	2	31.163	12.022	15.592	115.223	116.941
	3	49.556	28.649	54.685	517.123	138.111
	4	43.754	18.714	23.160	200.376	131.746
	5	14.199	5.572	10.493	97.739	43.224
	6	9.320	5.007	10.401	130.275	24.139
	7	8.566	3.855	6.114	79.467	31.069

*Categories of post high-school status: (1) enrolled in college (2) enrolled in non-collegiate school; (3) employed full-time; (4) other

$x^*_b(hijk)$

High-School Rank

Middle Third			Upper Third			
2	3	4	1	2	3	4
6.321	16.003	118.183	246.988	4.777	12.532	56.719
13.670	27.084	232.770	167.777	9.608	19.726	103.897
17.322	50.513	555.529	135.472	11.626	35.129	236.772
15.977	30.208	303.957	148.860	9.451	18.515	114.179
3.770	10.847	117.499	44.868	2.613	7.792	51.728
3.008	9.545	139.038	24.708	1.970	6.479	57.842
2.956	7.162	108.251	28.332	1.893	4.752	44.023
19.201	32.465	116.935	305.512	22.280	39.034	86.165
39.016	51.630	216.412	242.901	52.445	71.912	184.734
69.052	134.484	721.363	231.383	74.866	151.084	496.662
48.731	61.535	301.989	230.897	55.270	72.318	217.508
14.668	28.187	148.923	67.690	14.865	29.599	95.843
11.215	23.770	168.879	36.393	10.942	24.030	104.634
12.093	19.569	144.270	45.219	11.390	19.099	86.292

TABLE 4

Analysis of Information

	Component due to	Information	D.F.
a)	x(hij.), x(...k)	$2I(x:x_a^*) = 2824.434$	123
b)	x(hij.), x(h..k), x(.i.k),	$2I(x_b^*:x_a^*) = 2672.724$	27
	x(..jk)	$2I(x:x_b^*) = 151.710$	96
c)	x(hij.), x(.i.k), x(h.jk)	$2I(x_c^*:x_b^*) = 52.850$	18
		$2I(x:x_c^*) = 98.860$	78

Parametric Representations for Log-odds of x_b^*

$$\ell n\ x_b^*(hij1)/x_b^*(hij4) = \tau_1^k + \tau_{h1}^{hk} + \tau_{i1}^{ik} + \tau_{j1}^{jk},$$

$$\ell n\ x_b^*(hij2)/x_b^*(hij4) = \tau_2^k + \tau_{h2}^{hk} + \tau_{i2}^{ik} + \tau_{j2}^{jk},$$

$$\ell n\ x_b^*(hij3)/x_b^*(hij4) = \tau_3^k + \tau_{h3}^{hk} + \tau_{i3}^{ik} + \tau_{j3}^{jk}.$$

The values of the natural parameters in the log-odds representations are:

$\tau_1^k = -0.6462$	$\tau_2^k = -2.0250$	$\tau_3^k = -1.5081$
$\tau_{11}^{hk} = 0.2055$	$\tau_{12}^{hk} = -1.1216$	$\tau_{13}^{hk} = -0.7180$
$\tau_{11}^{ik} = -1.5814$	$\tau_{12}^{ik} = -1.0009$	$\tau_{13}^{ik} = -1.0566$
$\tau_{21}^{ik} = -0.8893$	$\tau_{22}^{ik} = -0.4541$	$\tau_{23}^{ik} = -0.4896$
$\tau_{11}^{jk} = 1.9119$	$\tau_{12}^{jk} = 0.6724$	$\tau_{13}^{jk} = 0.7163$
$\tau_{21}^{jk} = 0.9200$	$\tau_{22}^{jk} = 0.7658$	$\tau_{23}^{jk} = 0.5646$
$\tau_{31}^{jk} = -0.1176$	$\tau_{32}^{jk} = 0.1328$	$\tau_{33}^{jk} = 0.3180$
$\tau_{41}^{jk} = 0.7060$	$\tau_{42}^{jk} = 0.6550$	$\tau_{43}^{jk} = 0.4069$
$\tau_{51}^{jk} = 0.2984$	$\tau_{52}^{jk} = 0.1613$	$\tau_{53}^{jk} = 0.3331$
$\tau_{61}^{jk} = -0.4099$	$\tau_{62}^{jk} = -0.2329$	$\tau_{63}^{jk} = 0.0370$

EXAMPLE 5

Coronary heart disease risk. This example illustrates the analysis of a three-way 2 x 4 x 4 contingency table. It illustrates the test of equality of certain natural parameters in the model of no second-order interaction, both by computing the MDI estimate implied by the hypothesized relation among some of the natural parameters, and also by computing the appropriate quadratic approximation.

CORONARY HEART DISEASE RISK

We are indebted to Professor S. Greenhouse and J. Cornfield (1962) for calling our attention to this set of data.

In this example we analyze data from a 3-way, R x S x T, table resulting from a coronary heart disease study. We denote the observed values by x(ijk), where

Characteristic		Index	1	2	3	4
Coronary heart disease	R	i	yes	no		
Serum cholesterol, mg/100 cc	S	j	< 200	200-219	220-259	260 +
Blood pressure, mm Hg	T	k	< 127	127-146	147-166	167 +

The complete 2 x 4 x 4 table is given in Table 1. A preliminary analysis is given in the analysis of information Table 2, where the various sets of marginal constraints, the corresponding MDI statistics (information numbers), and degrees of freedom are listed. Interaction hypotheses corresponding to sets of marginal constraints in the table are

$$H_a: \ p(ijk) = p(i..)p(.jk)$$

$$H_b: \ p(ijk) = p(ij.)p(.jk)/p(.j.)$$

$$H_2: \ \text{no second-order interaction.}$$

The effects due to addition of each of the three two-way marginals are shown immediately above the measures of interaction. We note that both the information numbers and the degrees of freedom are additive.

This analysis indicated that a fit to this set of data could be made adequately using as explanatory variables the set of two-way marginals. The hypothesis tested was that of no second-order interaction in the sense of Bartlett (1935). See Ku, Varner, and Kullback (1971). We start with H_a because our first concern is whether the incidence of coronary heart disease is homogeneous over the factors serum cholesterol and blood pressure. Thus considering $2I(x:x_a^*)$ in Table 2 as the total unexplained variation we may set up the summary analysis of information Table 3.

The interpretation of the no second-order interaction hypothesis is:

a. The association between blood pressure and heart disease is the same for different levels of cholesterol,
b. The association between cholesterol level and heart disease is the same for different levels of blood pressure,
c. The association between cholesterol level and blood pressure is the same for subjects with and without heart disease.

For the MDI estimate x_2^* under the model of no second-order interaction the log-odds (logit) of the estimated incidence of coronary heart disease is a linear additive function of an average effect, an effect due to cholesterol and an effect due to blood pressure, i.e.,

$$\ln \frac{x_2^*(1jk)}{x_2^*(2jk)} = \tau_1^i + \tau_{1j}^{ij} + \tau_{1k}^{ik} .$$

Values of x_2^* are shown in Table 4 and the design matrix in Table 5.

There are 22 natural parameters, in addition to τ_0, to be estimated from the x_2^* values. A complete model would include nine additional natural parameters, which, under the no second-order interaction hypothesis, are equal to zero, i.e.,

$$\tau_{111}^{ijk} = \tau_{112}^{ijk} = \tau_{113}^{ijk} = 0,$$

$$\tau_{121}^{ijk} = \tau_{122}^{ijk} = \tau_{123}^{ijk} = 0,$$

$$\tau_{131}^{ijk} = \tau_{132}^{ijk} = \tau_{133}^{ijk} = 0.$$

The number of natural parameters in the complete model is 23 + 9 = 32, that is, the number of cells.

The computation of the natural parameter estimates is straightforward, e.g.,

$$\tau_1^i = \ln \frac{x_2^*(144)}{x_2^*(244)} = -0.9374,$$

etc. The values of the natural parameters (taus) are listed in Table 6.

When the dependent or response variable is dichotomous, odds and log-odds have long been used as indices indicative of risk. The estimated log-odds,

$$\ln \frac{x_2^*(1jk)}{x_2^*(2jk)} = \tau_1^i + \tau_{1j}^{ij} + \tau_{1k}^{ik},$$

and the estimated odds,

$$\frac{x_2^*(1jk)}{x_2^*(2jk)} = \exp(\tau_1^i)\exp(\tau_{1j}^{ij})\exp(\tau_{1k}^{ik})$$

are given in Table 7.

From the design matrix or the representation of the log-odds we can compute the difference in log-odds of risk of heart disease for change in blood pressure and constant cholesterol concentration in terms of the natural parameters (taus), e.g.,

$$\ln \frac{x_2^*(1j2)}{x_2^*(2j2)} - \ln \frac{x_2^*(1j1)}{x_2^*(2j1)} = \ln \frac{x_2^*(112)}{x_2^*(212)} - \ln \frac{x_2^*(111)}{x_2^*(211)}$$

$$= \tau_{12}^{ik} - \tau_{11}^{ik} = -0.0415.$$

Similarly,

$$\ln \frac{x_2^*(1j3)}{x_2^*(2j3)} - \ln \frac{x_2^*(1j2)}{x_2^*(2j2)} = 0.5738,$$

$$\ln \frac{x_2^*(1j4)}{x_2^*(2j4)} - \ln \frac{x_2^*(1j3)}{x_2^*(2j3)} = 0.6681.$$

The differences in log-odds for change in cholesterol level and constant blood pressure are:

$$\ln \frac{x_2^*(12k)}{x_2^*(22k)} - \ln \frac{x_2^*(11k)}{x_2^*(21k)} = -0.2079,$$

$$\ln \frac{x_2^*(13k)}{x_2^*(23k)} - \ln \frac{x_2^*(12k)}{x_2^*(22k)} = 0.7702,$$

$$\ln \frac{x_2^*(14k)}{x_2^*(24k)} - \ln \frac{x_2^*(13k)}{x_2^*(23k)} = 0.7818.$$

The differences in log-odds for change in cholesterol level and change in blood pressure are

$$\ln \frac{x_2^*(122)}{x_2^*(222)} - \ln \frac{x_2^*(111)}{x_2^*(211)} = -0.2494,$$

$$\ln \frac{x_2^*(133)}{x_2^*(233)} - \ln \frac{x_2^*(122)}{x_2^*(222)} = 1.3440,$$

$$\ln \frac{x_2^*(144)}{x_2^*(244)} - \ln \frac{x_2^*(133)}{x_2^*(233)} = 1.4499.$$

In view of the negative values of the changes in log-odds represented by $\tau_{12}^{ik} - \tau_{11}^{ik}$, $\tau_{12}^{ij} - \tau_{11}^{ij}$, we may wish to check the hypothesis that

$$\tau_{11}^{ij} = \tau_{12}^{ij}; \ \tau_{11}^{ik} = \tau_{12}^{ik} ,$$

which could imply that the risk does not begin to manifest itself significantly until the cholesterol level and blood pressure exceed some minimum level, that is, a threshold effect. Let

$$Z_1 = \tau_{12}^{ij} - \tau_{11}^{ij} = -0.2079$$

$$Z_2 = \tau_{12}^{ik} - \tau_{11}^{ik} = -0.0415.$$

The covariance matrix of the natural parameters (taus) for x_2^* is obtained as follows (a weighted version of Kullback (1959, 217)): Compute $\underset{\sim}{S} = \underset{\sim}{T}'\underset{\sim}{D}\underset{\sim}{T}$ where $\underset{\sim}{T}$ is the 32 x 23 design matrix for the loglinear representation of x_2^* in Table 5, $\underset{\sim}{D}$ is a diagonal matrix whose entries are the values of x_2^* in lexicographic order. Partition the matrix $\underset{\sim}{S}$ as

$$\begin{bmatrix} \underset{\sim}{S}_{11} & \underset{\sim}{S}_{12} \\ \underset{\sim}{S}_{21} & \underset{\sim}{S}_{22} \end{bmatrix}$$

where $\underset{\sim}{S}_{11}$ is 1 x 1.

Then the covariance matrix of the natural parameters (taus) is

$$(\underset{\sim}{S}_{22} - \underset{\sim}{S}_{21}\underset{\sim}{S}_{11}^{-1}\underset{\sim}{S}_{12})^{-1} \text{ or } \underset{\sim}{S}_{22.1}^{-1}.$$

The covariance matrix of Z_1, Z_2 is found to be:

$$a_{11} = s^{8,8} + s^{9,9} - 2s^{8,9} = 0.2175$$

$$a_{12} = a_{21} = s^{8,11} - s^{9,11} - s^{8,12} + s^{9,12} = -0.0013$$

$$a_{22} = s^{11,11} + s^{12,12} - 2s^{11,12} = 0.0922.$$

We found $A^{-1} = \begin{bmatrix} 4.5981 & 0.0648 \\ 0.0648 & 10.8469 \end{bmatrix}$,

$$X^2 = (Z_1, Z_2)\underset{\sim}{A}^{-1}\begin{pmatrix} Z_1 \\ Z_2 \end{pmatrix} = 0.2185$$

does not exceed the upper 5% critical value of a chi-square variate with 2 degrees of freedom.

For this particular hypothesis, we may alternatively revise the design matrix by combining the columns τ_{11}^{ij} with τ_{12}^{ij}, and τ_{11}^{ik} with τ_{12}^{ik}, and use the iterative procedure suggested by Gokhale (1972), Kullback (1973) for unusual marginal totals to obtain the MDI estimates of cell frequencies. The design matrix for the MDI estimate x_d^* is given in Table 5a. The MDI estimates x_d^* are given in Table 8. In Table 9 the log-odds $\ell n(x_d^*(1jk)/x_d^*(2jk))$ and the odds $x_d^*(1jk)/x_d^*(2jk)$ are listed. The multiplicative representations for the odds are given in Table 9a.

The associated analysis of information is Table 10. Note that $2I(x_2^*:x_d^*)$ tests the null hypothesis that $\tau_{11}^{ij} = \tau_{12}^{ij}$, $\tau_{11}^{ik} = \tau_{12}^{ik}$ and is approximated by the test previously given as a quadratic chi-square variate.

TABLE 1

Coronary Heart Disease Risk

	j: Serum cholesterol, mg/100 cc		k: blood pressure, mm Hg 1 < 127	2 127-146	3 147-166	4 167+	Total
	1	< 200	2	3	3	4	12
CHD	2	200-219	3	2	0	3	8
i = 1	3	220-259	8	11	6	6	31
	4	260+	7	12	11	11	41
	j total		20	28	20	24	92
	1	< 200	117	121	47	22	307
NCHD	2	200-219	85	98	43	20	246
i = 2	3	220-259	119	209	68	43	439
	4	260+	67	99	46	33	245
	j total		388	527	204	118	1237
	Total		408	555	224	142	1329

TABLE 2

Analysis of Information - Coronary Heart Disease Risk Data

Component due to	Information	D.F.
a) x(i..), x(.jk)		
Independence R x ST	$2I(x:x_a^*) = 58.726$	15
b) x(.jk), x(ij.)		
RS effect|ST	$2I(x_b^*:x_a^*) = 31.921$	3
Conditional independence R x T|S	$2I(x:x_b^*) = 26.805$	12
2) x(.jk), x(ij.), x(i.k)		
RT effect|ST, RS	$2I(x_2^*:x_b^*) = 18.730$	3
Second-order interaction	$2I(x:x_2^*) = 8.075$	9

TABLE 3

Analysis of Information

Component due to		Information	D.F.
x(i..), x(.jk),	Total	$2I(x:x_a^*) = 58.726$	15
x(.jk), x(ij.),	Cholesterol effect	$2I(x_b^*:x_a^*) = 31.921$	3
x(.jk), x(ij.), x(i.k),	Blood Pressure effect given Cholesterol	$2I(x_2^*:x_b^*) = 18.730$	3
Second-order interaction	(Residual)	$2I(x:x_2^*) = 8.075$	9

TABLE 4

Estimated Cell Frequencies under No Second-Order Interaction Hypothesis, x_2^*

	j:	Serum cholesterol, mg/100 cc	k: blood pressure, mm Hg 1 < 127	2 127-146	3 147-166	4 167 +	Total
	1	< 200	3.550	3.553	2.488	2.409	12.000
CHD	2	200-219	2.144	2.340	1.754	1.762	8.000
i = 1	3	220-259	6.501	10.827	6.227	7.446	31.001
	4	260 +	7.805	11.280	9.531	12.382	40.998
		Total	20.000	28.000	20.000	23.999	91.999
	1	< 200	115.450	120.447	47.512	23.591	307.000
NCHD	2	200-219	85.856	97.660	41.246	21.238	246.000
i = 2	3	220-259	120.499	209.173	67.773	41.554	438.999
	4	260 +	66.196	99.720	47.469	31.617	245.002
		Total	388.001	527.000	204.000	118.000	1237.001
		TOTAL	408.001	555.000	224.000	141.999	1329.000

TABLE 5

Design Matrix - Coronary Heart Disease Risk

				1	2	3	4	5	6	7	8	9	10	11	12	13	14	15	16	17	18	19	20	21	22
i	j	k	τ_0	i 1	j 1	j 2	j 3	k 1	k 2	k 3	ij 11	ij 12	ij 13	ik 11	ik 12	ik 13	jk 11	jk 12	jk 13	jk 21	jk 22	jk 23	jk 31	jk 32	jk 33
1	1	1	1	1	1			1			1			1			1								
1	1	2	1	1	1				1		1				1			1							
1	1	3	1	1	1					1	1					1			1						
1	1	4	1	1	1						1														
1	2	1	1	1		1		1				1		1						1					
1	2	2	1	1		1			1			1			1						1				
1	2	3	1	1		1				1		1				1						1			
1	2	4	1	1		1						1													
1	3	1	1	1			1	1					1	1									1		
1	3	2	1	1			1		1				1		1									1	
1	3	3	1	1			1			1			1			1									1
1	3	4	1	1			1						1												
1	4	1	1	1				1						1											
1	4	2	1	1					1						1										
1	4	3	1	1						1						1									
1	4	4	1	1																					
2	1	1	1		1			1									1								
2	1	2	1		1				1									1							
2	1	3	1		1					1									1						
2	1	4	1		1																				
2	2	1	1			1		1												1					
2	2	2	1			1			1												1				
2	2	3	1			1				1												1			
2	2	4	1			1																			
2	3	1	1				1	1															1		
2	3	2	1				1		1															1	
2	3	3	1				1			1															1
2	3	4	1				1																		
2	4	1	1					1																	
2	4	2	1						1																
2	4	3	1							1															
2	4	4	1																						

TABLE 5a

Design Matrix For MDI Estimate x_d^*

		1	2	3	4	5	6	7	8	9	10	11	12	13	14	15	16	17	18	19	20
	τ_0	i	j	j	j	k	k	k	ij	ij	ik	ik	jk	jk	jk	jk	jk	jk	jk	jk	jk
									11		11										
i j k		1	1	2	3	1	2	3	12	13	12	13	11	12	13	21	22	23	31	32	33
1 1 1	1	1	1			1			1		1		1								
1 1 2	1	1	1				1		1		1			1							
1 1 3	1	1	1					1	1			1			1						
1 1 4	1	1	1						1												
1 2 1	1	1		1		1			1		1					1					
1 2 2	1	1		1			1		1		1						1				
1 2 3	1	1		1				1	1			1						1			
1 2 4	1	1		1					1												
1 3 1	1	1			1	1				1	1								1		
1 3 2	1	1			1		1			1	1									1	
1 3 3	1	1			1			1		1		1									1
1 3 4	1	1			1					1											
1 4 1	1	1				1					1										
1 4 2	1	1					1				1										
1 4 3	1	1						1				1									
1 4 4	1	1																			
2 1 1	1		1			1							1								
2 1 2	1		1				1							1							
2 1 3	1		1					1							1						
2 1 4	1		1																		
2 2 1	1			1		1										1					
2 2 2	1			1			1										1				
2 2 3	1			1				1										1			
2 2 4	1			1																	
2 3 1	1				1	1													1		
2 3 2	1				1		1													1	
2 3 3	1				1			1													1
2 3 4	1				1																
2 4 1	1					1															
2 4 2	1						1														
2 4 3	1							1													
2 4 4	1																				

TABLE 6

Estimates of the Natural Parameters, x_2^*

$\tau_1^i = -0.9374$	$\tau_{11}^{ij} = -1.3441$	$\tau_{11}^{jk} = 0.8491$
$\tau_1^j = -0.2929$	$\tau_{12}^{ij} = -1.5520$	$\tau_{12}^{jk} = 0.4817$
$\tau_2^j = -0.3979$	$\tau_{13}^{ij} = -0.7818$	$\tau_{13}^{jk} = 0.2938$
$\tau_3^j = 0.2733$	$\tau_{11}^{ik} = -1.2004$	$\tau_{21}^{jk} = 0.6580$
$\tau_1^k = 0.7389$	$\tau_{12}^{ik} = -1.2419$	$\tau_{22}^{jk} = 0.3770$
$\tau_2^k = 1.1481$	$\tau_{13}^{ik} = -0.6681$	$\tau_{23}^{jk} = 0.2574$
$\tau_3^k = 0.4064$		$\tau_{31}^{jk} = 0.3527$
		$\tau_{32}^{jk} = 0.4675$
		$\tau_{33}^{jk} = 0.0828$

NOTE: Any natural parameter corresponding to a subscript $i = 2$, and/or $j = 4$, and/or $k = 4$ is zero by convention.

TABLE 7

Log-odds and Odds, x_2^*

Entries are

$$\text{log-odds } \ln \frac{x_2^*(1jk)}{x_2^*(2jk)} \text{ and odds } \frac{x_2^*(1jk)}{x_2^*(2jk)}$$

	k = 1	k = 2	k = 3	k = 4
j = 1	- 3.482	- 3.523	- 2.950	- 2.281
	.0307	.0295	.0523	.1022
j = 2	- 3.690	- 3.731	- 3.158	- 2.489
	.0250	.0240	.0245	.0830
j = 3	- 2.920	- 2.961	- 2.387	- 1.719
	.0539	.0518	.0919	.1792
j = 4	- 2.138	- 2.179	- 1.605	- 0.937
	0.1179	0.1132	0.2009	0.3918

TABLE 8

MDI Estimate, x_d^*

	j: Serum cholesterol, mg/100 cc		k: blood pressure, mm Hg 1 < 127	2 127-146	3 147-166	4 167 +	Total
	1	< 200	3.189	3.323	2.289	2.225	11.026
CHD	2	200-219	2.358	2.680	1.969	1.968	8.975
i = 1	3	220-259	6.350	11.000	6.217	7.434	31.001
	4	260 +	7.640	11.460	9.525	12.374	40.999
			19.537	28.463	20.000	24.001	92.001
	1	< 200	115.811	120.677	47.711	23.775	307.974
NCHD	2	200-219	85.642	97.320	41.031	21.032	245.025
i = 2	3	220-259	120.650	209.000	67.783	41.566	438.999
	4	260 +	66.360	99.539	47.475	31.626	245.000
		Total	388.463	526.536	204.000	117.999	1236.998
		TOTAL	408.000	554.999	224.000	142.000	1328.999

TABLE 9

Log-odds, and Odds, MDI Estimate x_d^*

		Blood pressure			
		k = 1	2	3	4
	j = 1	- 3.592	- 3.592	- 3.037	- 2.369
		0.0275	0.0275	0.0480	0.0936
	j = 2	- 3.592	- 3.592	- 3.037	- 2.369
Serum cholesterol		0.0275	0.0275	0.0480	0.0936
	j = 3	- 2.944	- 2.944	- 2.389	- 1.721
		0.0526	0.0526	0.0917	0.1788
	j = 4	- 2.162	- 2.162	- 1.606	- 0.938
		0.1151	0.1151	0.2006	0.3912

$$\ell n\ x_d^*(1jk)/x_d^*(2jk) = \tau_1^i + \tau_{1j}^{ij} + \tau_{1k}^{ik}$$

$$\tau_1^i = -0.9384$$

$$\tau_{11}^{ij} = \tau_{12}^{ij} = -1.4306 \qquad \tau_{11}^{ik} = \tau_{12}^{ik} = -1.2232$$

$$\tau_{13}^{ij} = -0.7828 \qquad \tau_{13}^{ik} = -0.6678$$

TABLE 9a

Multiplicative Representation

		Blood Pressure	
		k = 1	2
	j = 1	.3913 x .2392 x .2943	.3913 x .2392 x .2943
Serum	j = 2	.3913 x .2392 x .2943	.3913 x .2392 x .2943
cholesterol	j = 3	.3913 x .4571 x .2943	.3913 x .4571 x .2943
	j = 4	.3913 x 1 x .2943	.3913 x 1 x .2943

ODDS

$$\frac{x_d^*(ijk)}{x_d^*(2jk)} = \exp(\tau_1^i)\ \exp(\tau_{1j}^{ij})\ \exp(\tau_{1k}^{ik})$$

$\exp(\tau_1^i) = 0.3913$, $\exp(\tau_{11}^{ij}) = \exp(\tau_{12}^{ij}) = 0.2392$, $\exp(\tau_{13}^{ij}) = 0.4571$

$\exp(\tau_{14}^{ij}) = 1$, $\exp(\tau_{11}^{ik}) = \exp(\tau_{12}^{ik}) = 0.2943$, $\exp(\tau_{13}^{ik}) = 0.5128$

$\exp(\tau_{14}^{ik}) = 1.$

3	4
.3913 x .2392 x .5128	.3913 x .2392 x 1
.3913 x .2392 x .5128	.3913 x .2392 x 1
.3913 x .4571 x .5128	.3913 x .4571 x 1
.3913 x 1 x .5128	.3913 x 1 x 1

TABLE 10

Analysis of Information

	Component due to	Information	D.F.
a)	x(i..), x(.jk)	$2I(x:x^*_a) = 58.726$	15
d)	x(.jk), x(ij.), j = 3,4 x(i.k), k = 3,4	$2I(x^*_d:x^*_a) = 50.429$	4
	x(i1.) + x(i2.); x(i.1) + x(i.2)	$2I(x:x^*_d) = 8.297$	11
2)	x(.jk), x(ij.), x(i.k)	$2I(x^*_2:x^*_d) = 0.222$	2
		$2I(x:x^*_2) = 8.075$	9

EXAMPLE 6

Hospital data. This example illustrates the analysis of a pair of related three-way 2 x 2 x 2 contingency tables. In particular it illustrates the procedure to obtain an MDI estimate satisfying certain observed marginal constraints and having certain of the natural parameters (taus) predetermined, that is, the inheritance of certain natural parameters. It also mentions that the T-functions of the two-way marginals are the products of the T-functions of the related one-way marginals. The related 2 x 2 x 2 tables were also examined as a four-way 2 x 2 x 2 x 2 table to compare the results and conclusions.

HOSPITAL DATA

The data used are from the field of hospital administration and relate to the matter of innovation in hospitals. We begin with the assumption that the use of electronic data processing (EDP) in hospitals in the late 1960's was innovative. This assumption is substantiated by a variety of surveys of the use of EDP in hospitals (Hammon, Jacobs, and Reeves, 1972). On this basis the data in a survey of hospitals using EDP conducted by Herner and Co. were combined with data from the <u>Guide Issue of</u>

Hospitals for the same period so that a file of records reflecting characteristics of hospitals and levels at which EDP was used by these hospitals was created. The hospitals in this survey were selected by stratified sampling. The stratification (fixed variable) was on the basis of hospital size. All hospitals in the large-size category (200 or more beds) were included in the survey and a ten percent sample was taken of those in the small size category. The data from these files were tabulated and arranged in multiway contingency tables. The analysis of the tables for the large and small hospitals will be described here and interrelated.

On the basis of these analyses we conclude that there is a distinct relation of innovation on location and length of stay with a common factor for large and small hospitals. The association (measured by the logarithm of the cross-product ratio) between use of EDP and length of stay is the same for the large and small hospitals. The log-odds (logit) of use of EDP in descending order of magnitude within the large hospitals and within the small hospitals are parallel in terms of the combinations of the factors location and length of stay. The usage of EDP is generally greater in the large hospitals than in the small hospitals except that the best log-odds for the small hospitals is greater than the poorest log-odds for the large hospitals.

In a study to identify characteristics which distinguish hospitals which use EDP from those which do not, that is, to identify characteristics which are significantly associated with use of EDP, data on 1176 hospitals, 923 large and 253 small, were collected with respect to use, location, and length of stay. The data appear in the two three-way 2 x 2 x 2 contingency Tables 1 and 2. In order to determine the relation among the free variables use, location and length of stay, indexed by size of hospital, and interactions that may exist among these characteristics it seems intuitively clear that an analysis based only on two-way tables would not suffice.

We shall denote the occurrences in the observed Tables 1 and 2 respectively by x(ijk), y(ijk) where

Characteristic	Index	1	2
Usage	i	User	Non-user
Location	j	Urban	Rural
Stay	k	Short	Long

The proposed procedure provides MDI estimates for the original data analogous to a regression procedure using sets of observed marginals as explanatory variables. We shall try to find MDI estimates which do not differ significantly from the observed data. The pair of acceptable MDI estimates will indicate the nature of the significant interactions for which we can compute numerical measures.

As a first step in the analysis we shall find smoothed estimates of the original data. We shall do this for the large hospitals also even though the data for all large hospitals was collected. We examine the MDI estimates obtained by successively fitting sets of observed marginals. It turns out that the sets of two-way marginals are best and the resultant MDI estimates provide a satisfactory fit. The estimated tables have the same two-way and also the same one-way marginals as the original tables. These MDI estimates which we denote by $x_2^*(ijk)$, $y_2^*(ijk)$ respectively for the large and small hospitals are given in Tables 3 and 4 and imply no second-order (three-factor) interaction. Note that the MDI estimate for the observed value y(122) = 0 is $y_2^*(122) = 0.137$.

The MDI estimates are given analytically by the loglinear representation

$$\ell n \frac{x_2^*(ijk)}{n\pi(ijk)} = L + \tau_1^i T_1^i(ijk) + \tau_1^j T_1^j(ijk) + \tau_1^k T_1^k(ijk) \qquad (1)$$

$$+ \tau_{11}^{ij} T_{11}^{ij}(ijk) + \tau_{11}^{ik} T_{11}^{ik}(ijk) + \tau_{11}^{jk} T_{11}^{jk}(ijk)$$

where $n = \Sigma\Sigma\Sigma x(ijk)$, $\pi(ijk) = 1/8$, L is a normalizing constant, the natural parameters (taus) are main-effect and interaction parameters, and the T(ijk) are a set of linearly independent random variables, in this case the indicator functions of the respective marginals. A similar representation holds for $y_2^*(ijk)$. The loglinear representations are shown graphically in Table 6. The zeros or ones in the various columns of Table 6 are the values of the respective functions T(ijk). Note that

$$T_{11}^{ij}(ijk) = T_1^i(ijk)T_1^j(ijk),\quad T_{11}^{ik}(ijk) = T_1^i(ijk)T_1^k(ijk),$$

$$T_{11}^{jk}(ijk) = T_1^j(ijk)T_1^k(ijk).$$

To test the goodness-of-fit of the MDI estimates we compute

$$2I(x:x_2^*) = 2\Sigma\Sigma\Sigma x(ijk)\ell n(x(ijk)/x_2^*(ijk)) = 0.481,\ 1\text{ D.F.}$$

$$2I(y:y_2^*) = 2\Sigma\Sigma\Sigma y(ijk)\ell n(y(ijk)/y_2^*(ijk)) = 0.294,\ 1\text{ D.F.}$$

Since the MDI statistics are asymptotically distributed as chi-square, we conclude that the smoothed values x_2^*, y_2^* are good estimates and we shall use them in our subsequent analysis.

From the loglinear representation (1) or the graphical presentation in Table 6, we find that the log-odds or logits of the use of EDP for large hospitals is given by the parametric representation

$$\ell n \frac{x_2^*(111)}{x_2^*(211)} = \tau_1^i + \tau_{11}^{ij} + \tau_{11}^{ik}$$

(2)

$$\ell n \frac{x_2^*(112)}{x_2^*(212)} = \tau_1^i + \tau_{11}^{ij}$$

$$\ell n \frac{x_2^*(121)}{x_2^*(221)} = \tau_1^i + \tau_{11}^{ik}$$

$$\ell n \frac{x_2^*(122)}{x_2^*(222)} = \tau_1^i$$

where the natural parameters for the MDI estimate $x_2^*(ijk)$ are found to be

$$\tau_1^i = -1.4842, \ \tau_{11}^{ij} = 0.5113, \ \tau_{11}^{ik} = 1.5103.$$

From (2) we also see that for the large hospitals

$$\tau_{11}^{ij} = \ln \frac{x_2^*(111)x_2^*(221)}{x_2^*(211)x_2^*(121)} = \ln \frac{x_2^*(112)x_2^*(222)}{x_2^*(212)x_2^*(122)} = 0.5113,$$

that is, the association between usage and location for either short or long stay. Similarly

$$\tau_{11}^{ik} = \ln \frac{x_2^*(111)x_2^*(212)}{x_2^*(211)x_2^*(112)} = \ln \frac{x_2^*(121)x_2^*(222)}{x_2^*(221)x_2^*(122)} = 1.5103,$$

that is, the association between usage and stay for either urban or rural location.

For the small hospitals the log-odds or logits are

$$\ln \frac{y_2^*(111)}{y_2^*(211)} = \tau_1^i + \tau_{11}^{ij} + \tau_{11}^{ik}$$

$$\ln \frac{y_2^*(112)}{y_2^*(212)} = \tau_1^i + \tau_{11}^{ij}$$

$$\ln \frac{y_2^*(121)}{y_2^*(221)} = \tau_1^i + \tau_{11}^{ik}$$

$$\ln \frac{y_2^*(122)}{y_2^*(222)} = \tau_1^i$$

where the natural parameters for the MDI estimate $y_2^*(ijk)$ are found to be

$$\tau_1^i = -3.3357,\ \tau_{11}^{ij} = 1.3088,\ \tau_{11}^{ik} = 0.9836.$$

For the small hospitals we also have

$$\tau_{11}^{ij} = \ln \frac{y_2^*(111)y_2^*(221)}{y_2^*(211)y_2^*(121)} = \ln \frac{y_2^*(112)y_2^*(222)}{y_2^*(212)y_2^*(122)} = 1.3088,$$

that is, the association between usage and location for either short or long stay. Similarly

$$\tau_{11}^{ik} = \ln \frac{y_2^*(111)y_2^*(212)}{y_2^*(211)y_2^*(112)} = \ln \frac{y_2^*(121)y_2^*(222)}{y_2^*(221)y_2^*(122)} = 0.9836,$$

that is the association between usage and stay for either urban or rural locations.

Since the data for the large hospitals reflect observations over all such hospitals, it will be of interest to determine whether there exists a suitable MDI estimate for the small hospitals, other than $y_2^*(ijk)$, with some of its natural parameters the same as the corresponding values for the large hospitals. This can be accomplished by using the iterative algorithm fitting various subsets of marginals of $y_2^*(ijk)$ (or the original $y(ijk)$) but starting the iteration with a distribution which has the same natural parameters (taus) as $x_2^*(ijk)$. The natural parameters (taus) of $x_2^*(ijk)$ not affected by the marginal fitting constraints will be inherited by the resultant MDI estimate. We shall use $v(ijk) = (253/923)x_2^*(ijk)$ which has the same natural parameters (taus) as $x_2^*(ijk)$ with total adjusted to be the same as the observed total of small hospitals.

We summarize the procedure: starting the iterative fitting algorithm with v(ijk) (recall that y(ijk) and $y_2^*(ijk)$ have the same two-way and one-way marginals)

	Marginals fitted	MDI Estimate	Natural parameters inherited from v(ijk)
a)	y(i.k),y(.jk)	$u_a^*(ijk)$	τ_{11}^{ij}
b)	y(ij.),y(.jk)	$u_b^*(ijk)$	τ_{11}^{ik}
c)	y(ij.),y(i.k)	$u_c^*(ijk)$	τ_{11}^{jk}
d)	y(.jk),y(i..)	$u_d^*(ijk)$	$\tau_{11}^{ij},\tau_{11}^{ik}$
e)	y(i.k),y(.j.)	$u_e^*(ijk)$	$\tau_{11}^{ij},\tau_{11}^{jk}$
f)	y(ij.),y(..k)	$u_f^*(ijk)$	$\tau_{11}^{ik},\tau_{11}^{jk}$
g)	y(i..),y(.j.),y(..k)	$u_g^*(ijk)$	$\tau_{11}^{ij},\tau_{11}^{ik},\tau_{11}^{jk}$

In order to test whether the u* estimates differ significantly from the y_2^* estimates, that is, whether the natural parameters in y_2^* differ significantly from the natural parameters in u* inherited from x_2^* or v, we compute

$$2I(y_2^*:u_m^*) = 2\Sigma\Sigma\Sigma y_2^*(ijk)\ln(y_2^*(ijk)/u_m^*(ijk))$$

which is asymptotically distributed as chi-square with 1 D.F. for m = a,b,c, 2 D.F. for m = d,e,f, 3 D.F. for m = g.

The only case which yielded a non-significant value was $u_b^*(ijk)$ for which

$$2I(y_2^*:u_b^*) = 0.408,\ 1\ \text{D.F.}$$

The values of $u_b^*(ijk)$ are given in Table 5. The loglinear representation for $u_b^*(ijk)$ in terms of $v(ijk)$ is

$$\ell n \frac{u_b^*(ijk)}{v(ijk)} = L + \tau_1^i T_1^i(ijk) + \tau_1^j T_1^j(ijk) + \tau_1^k T_1^k(ijk) + \tau_{11}^{ij} T_{11}^{ij}(ijk) + \tau_{11}^{jk} T_{11}^{jk}(ijk). \tag{3}$$

Note that τ_{11}^{ik} does not appear explicitly in (3). By using the loglinear representation for $v(ijk)$ itself we also get the reparametrization or loglinear representation for $u_b^*(ijk)$ in terms of the uniform distribution

$$\ell n \frac{u_b^*(ijk)}{n\pi(ijk)} = L + \tau_1^i T_1^i(ijk) + \tau_1^j T_1^j(ijk) + \tau_1^k T_1^k(ijk) + \tau_{11}^{ij} T_{11}^{ij}(ijk) + \tau_{11}^{ik} T_{11}^{ik}(ijk) + \tau_{11}^{jk} T_{11}^{jk}(ijk) \tag{4}$$

We remark that the numerical values of the taus in (3) and (4) are not the same.

The log-odds or logits of the use of EDP for small hospitals may now be given by the parametric representation

$$\ell n \frac{u_b^*(111)}{u_b^*(211)} = \tau_1^i + \tau_{11}^{ij} + \tau_{11}^{ik}$$

$$\ell n \frac{u_b^*(112)}{u_b^*(212)} = \tau_1^i + \tau_{11}^{ij}$$

(5)

$$\ell n \frac{u_b^*(121)}{u_b^*(221)} = \tau_1^i + \tau_{11}^{ik}$$

$$\ell n \frac{u_b^*(122)}{u_b^*(222)} = \tau_1^i$$

where the values of the natural parameters in (5) are

$$\tau_1^i = -3.8569, \ \tau_{11}^{ij} = 1.3354, \ \tau_{11}^{ik} = 1.5103.$$

For the small hospitals we now have the associations

$$\tau_{11}^{ij} = \ell n \frac{u_b^*(111)u_b^*(221)}{u_b^*(211)u_b^*(121)} = \ell n \frac{u_b^*(112)u_b^*(222)}{u_b^*(212)u_b^*(122)} = 1.3354$$

and

$$\tau_{11}^{ik} = \ell n \frac{u_b^*(111)u_b^*(212)}{u_b^*(211)u_b^*(112)} = \ell n \frac{u_b^*(121)u_b^*(222)}{u_b^*(221)u_b^*(122)} = 1.5103.$$

Note that 1.3354, the association between usage and location for the small hospitals is still different from that for the large hospitals, but that the association between usage and stay, 1.5103, is now the same for both large and small hospitals.

Arranging the log-odds of usage in descending order of magnitude within the large hospitals and within the small hospitals we find

Large hospitals	Factors	Small hospitals
$\ln \frac{x_2^*(111)}{x_2^*(211)} = 0.5374$	Urban,Short	$\ln \frac{u_b^*(111)}{u_b^*(211)} = -1.0111$
$\ln \frac{x_2^*(121)}{x_2^*(221)} = 0.0262$	Rural,Short	$\ln \frac{u_b^*(121)}{u_b^*(221)} = -2.3466$
$\ln \frac{x_2^*(112)}{x_2^*(212)} = -0.9729$	Urban,Long	$\ln \frac{u_b^*(112)}{u_b^*(212)} = -2.5214$
$\ln \frac{x_2^*(122)}{x_2^*(222)} = -1.4841$	Rural,Long	$\ln \frac{u_b^*(122)}{u_b^*(222)} = -3.8569$

It was considered of interest to assess the influence of the sampling procedure by comparing the results of the analysis of the hospital data as two separate three-way tables, with an analysis of the same data set up as the four-way 2 x 2 x 2 x 2 Table 7. (See Section V of Chapter 3.) We denote the occurrences in Table 7 by $x(ijk\ell)$, where

Characteristic	Index	1	2
Usage	i	User	Non-user
Location	j	Urban	Rural
Stay	k	Short	Long
Size	ℓ	Large	Small

A sequence of sets of marginals was fitted to provide an overall analysis of information in order to select an appropriate model for study. We omit details and present in the analysis of information table the results for the basic model and the final model $x_c^*(ijk\ell)$. The MDI estimates $x_c^*(ijk\ell)$ are given in Table 8. The representation for the log-odds is shown in equation (6) and the values of the natural parameters are given in equation (7).

Analysis of Information

Component due to	Information	D.F.
a) x(.jkℓ),x(i...)	$2I(x:x_a^*) = 231.461$	7
c) x(.jkℓ),x(i.k.),x(ij.ℓ)	$2I(x_c^*:x_a^*) = 230.296$	4
	$2I(x:x_c^*) = 1.165$	3

(6)

$$\ln \frac{x_c^*(1111)}{x_c^*(2111)} = \tau_1^i + \tau_{11}^{ij} + \tau_{11}^{ik} + \tau_{11}^{i\ell} + \tau_{111}^{ij\ell}$$

$$\ln \frac{x_c^*(1121)}{x_c^*(2121)} = \tau_1^i + \tau_{11}^{ij} + \tau_{11}^{i\ell} + \tau_{111}^{ij\ell}$$

$$\ln \frac{x_c^*(1211)}{x_c^*(2211)} = \tau_1^i + \tau_{11}^{ik} + \tau_{11}^{i\ell}$$

$$\ln \frac{x_c^*(1221)}{x_c^*(2221)} = \tau_1^i + \tau_{11}^{i\ell}$$

$$\ln \frac{x_c^*(1112)}{x_c^*(2112)} = \tau_1^i + \tau_{11}^{ij} + \tau_{11}^{ik}$$

$$\ln \frac{x_c^*(1122)}{x_c^*(2122)} = \tau_1^i + \tau_{11}^{ij}$$

$$\ln \frac{x_c^*(1212)}{x_c^*(2212)} = \tau_1^i + \tau_{11}^{ik}$$

$$\ln \frac{x_c^*(1222)}{x_c^*(2222)} = \tau_1^i$$

(7)

$$\tau_1^i = -3.8351, \quad \tau_{11}^{ij} = 1.3345, \quad \tau_{11}^{ik} = 1.4886$$

$$\tau_{11}^{i\ell} = 2.3657, \quad {}_{111}^{ij\ell} = -0.8203$$

From equation (6) it is found that the association between usage and stay is the same for all combinations of the other factors, that is,

$$\ell n \frac{x_c^*(1j1\ell)x_c^*(2j2\ell)}{x_c^*(2j1\ell)x_c^*(1j2\ell)} = \tau_{11}^{ik} = 1.4886.$$

This agrees with the conclusion from the previous analysis with a common value of 1.5103 for the association between usage and stay for the large and small hospitals.

The association between usage and location is given by

$$\ell n \frac{x_c^*(11k\ell)x_c^*(22k\ell)}{x_c^*(21k\ell)x_c^*(12k\ell)} = \tau_{11}^{ij} + \tau_{111}^{ij\ell},$$

that is, it varies with size but not with stay. Thus for the large hospitals the value is

$$\tau_{11}^{ij} + \tau_{111}^{ij\ell} = 1.3345 - 0.8203 = 0.5142,$$

and for the small hospitals it is

$$\tau_{11}^{ij} = 1.3345.$$

This is consistent with the previous conclusion with the values 0.5113, 1.3354 respectively for large and small hospitals.

Arranging the log-odds of usage in descending order of magnitude within the large hospitals and within the small hospitals we find

Large hospitals	Factors	Small hospitals
$\ell n \frac{x_c^*(1111)}{x_c^*(2111)} = 0.5334$	Urban, Short	$\ell n \frac{x_c^*(1112)}{x_c^*(2112)} = -1.0120$
$\ell n \frac{x_c^*(1211)}{x_c^*(2211)} = 0.0192$	Rural, Short	$\ell n \frac{x_c^*(1212)}{x_c^*(2212)} = -2.3465$
$\ell n \frac{x_c^*(1121)}{x_c^*(2121)} = -0.9552$	Urban, Long	$\ell n \frac{x_c^*(1122)}{x_c^*(2122)} = -2.5006$

$$\ln \frac{x_c^*(1221)}{x_c^*(2221)} = -1.4694 \qquad \text{Rural, Long} \qquad \ln \frac{x_c^*(1222)}{x_c^*(2222)} = -3.8351$$

We note that the results are consistent with those previously determined.

We leave it as an exercise for the reader to set up the odds factor tables for the two three-way and the four-way tables.

There seems to be no essential difference in conclusions about the data analyzed as two three-way tables or one four-way table.

TABLE 1

Large Hospitals x(ijk)

	Urban		Rural		
	Short	Long	Short	Long	
User	376	40	52	15	483
Non-user	217	112	54	57	440
	593	152	106	72	923

TABLE 2

Small Hospitals y(ijk)

	Urban		Rural		
	Short	Long	Short	Long	
User	28	2	11	0	41
Non-user	80	14	114	4	212
	108	16	125	4	253

TABLE 3

Large Hospitals $x_2^*(ijk)$

	Urban		Rural		
	Short	Long	Short	Long	
User	374.305	41.694	53.695	13.306	483.000
Non-user	218.693	110.308	52.307	58.692	440.000
	592.998	152.002	106.002	71.998	923.000

TABLE 4

Small Hospitals $y_2^*(ijk)$

	Urban		Rural		
	Short	Long	Short	Long	
User	28.137	1.863	10.863	0.137	41.000
Non-user	79.863	14.137	114.137	3.863	212.000
	108.000	16.000	125.000	4.000	253.000

TABLE 5

Small Hospitals $u_b^*(ijk)$

	Urban		Rural		
	Short	Long	Short	Long	
User	28.810	1.190	10.917	0.083	41.000
Non-user	79.190	14.810	114.083	3.917	212.000
	108.000	16.000	125.000	4.000	253.000

TABLE 6

Loglinear Representation

ijk	L	τ_1^i	τ_1^j	τ_1^k	τ_{11}^{ij}	τ_{11}^{ik}	τ_{11}^{jk}
111	1	1	1	1	1	1	1
112	1	1	1		1		
121	1	1		1		1	
122	1	1					
211	1		1	1			1
212	1		1				
221	1			1			
222	1						

TABLE 7

$x(ijk\ell)$

	Large				Small			
	Urban		Rural		Urban		Rural	
	Short	Long	Short	Long	Short	Long	Short	Long
User	376	40	52	15	28	2	11	0
Non-user	217	112	54	57	80	14	114	4

TABLE 8

$x_c^*(ijk\ell)$

	Large				Small			
	Urban		Rural		Urban		Rural	
	Short	Long	Short	Long	Short	Long	Short	Long
User	373.760	42.242	53.536	13.460	28.787	1.214	10.916	.085
Non-user	219.228	109.782	52.516	58.502	79.194	14.798	114.062	3.918

EXAMPLE 7

Partitioning using OUTLIERS. In Chapter 3 the notion of OUTLIER and its possible use was discussed. In this example we shall use the OUTLIER property to partition a 2 x 7 table into homogeneous segments. The data were discussed by Sugiura and Otake (1973) who reached the same partitioning by a different approach. Additional illustrations of the use of OUTLIER will be found in other examples.

PARTITIONING USING OUTLIERS

Table 1a presents data on leukemia cases observed. Denoting the entries in the observed table by $x(ij)$, $i = 1,2$, $j = 1,2, \ldots ,7$ we first test whether the incidence of leukemia is homogeneous over the doses by fitting the marginals $x(i.)$, $x(.j)$. We find that large OUTLIER values are associated with values of $j = 1,2,6,7$ and that $2I(x:x^*) = 44.65$, 6 D.F.

Since the doses are arranged on a scale we repeat the process omitting the cells $x(ij)$, $i = 1,2$, $j = 6,7$. We find that a large OUTLIER value is associated with $j = 3$ and that $2I(x:x^*) = 18.92$, 4 D.F.

We continue the process using the original cells corresponding to $j = 3,4,5$. Now we find no large OUTLIER values and $2I(x:x^*) = 0.09$, 2 D.F. For the original cells with $j = 6,7$ again there are no large OUTLIERS and $2I(x:x^*) = 0.37$, 1 D.F. For the original cells with $j = 1,2$ again there are no large OUTLIERS and $2I(x:x^*) = 0.91$, 1 D.F.

We summarize in the Analysis of Information Tables.

Component due to	Information	D.F.
cells j = 1, ... ,7	$2I(x:x^*) = 44.649$	6
omit cells j = 6,7	$2I(x_a^*:x^*) = 25.734$	2
cells j = 1, ... ,5[1]	$2I(x:x_a^*) = 18.915$	4
omit cells, j = 1,2	$2I(x_b^*:x_a^*) = 18.826$	2
cells j = 3,4,5[2]	$2I(x:x_b^*) = 0.089$	2
	$2I(x:x^*) = 44.649$	6
omit cells, j = 1,2,3,4,5	$2I(x_c^*:x^*) = 44.283$	5
cells j = 6,7[3]	$2I(x:x_c^*) = 0.366$	1
	$2I(x:x^*) = 44.649$	6
omit cells j = 3,4,5,6,7	$2I(x_d^*:x^*) = 43.740$	5
cell j = 1,2[4]	$2I(x:x_d^*) = 0.909$	1

[1]Note that $x_a^*(ij) = x(i.)x(.j)/n$, $i = 1,2$, $j = 1,2, \ldots ,5$

$x_a^*(ij) = x(ij)$, $i = 1,2$, $j = 6,7$

[2]Note that $x_b^*(ij) = x(i.)x(.j)/n$, $i = 1,2$, $j = 3,4,5$

$x_b^*(ij) = x(ij)$, $i = 1,2$, $j = 1,2,6,7$

[3]Note that $x_c^*(ij) = x(i.)x(.j)/n$, $i = 1,2$, $j = 6,7$

$x_c^*(ij) = x(ij)$, $i = 1,2$, $j = 1,2,3,4,5$

[4]Note that $x_d^*(ij) = x(i.)x(.j)/n$, $i = 1,2$, $j = 1,2$,

$x_d^*(ij) = x(ij)$, $i = 1,2$, $j = 3,4,5,6,7$

We now define an overall MDI estimate by

$$x_e^*(ij) = x_d^*(ij), \quad i = 1,2, \; j = 1,2$$

$$x_e^*(ij) = x_b^*(ij), \quad i = 1,2, \; j = 3,4,5$$

$$x_e^*(ij) = x_c^*(ij), \quad i = 1,2, \; j = 6,7$$

and we have for the associated MDI statistic

$$2I(x:x_e^*) = 1.364, \text{ 4 D.F.}$$

The values of $x_e^*(ij)$ are given in Table 1b.

TABLE 1a

Number of Leukemia Cases Observed for the Period
1 Oct 1950 - 30 Sept 1966 Among Hiroshima Male
Survivors for the Extended Life Span Study
Sample at Atomic Bomb Casualty Commission
Aged 15-19 at the Time of Atomic Bomb

Dose (rad)	< 5	5	20	50	100	200	300 +	Total
Leukemia	2	0	3	2	2	2	5	16
Not Leukemia	4601	1161	477	271	243	98	149	7000
Total	4603	1161	480	273	245	100	154	7016

TABLE 1b

Values of estimate $x_e^*(ij)$

Dose (rad)	< 5	5	20	50
Leukemia	1.597	0.403	3.367	1.915
Not Leukemia	4601.398	1160.597	476.633	271.085
	4602.995	1161.000	480.000	273.000

Dose (rad)	100	200	300 +	Total
Leukemia	1.718	2.756	4.244	16.000
Not Leukemia	243.282	97.244	149.756	6999.995
	245.000	100.000	154.000	7015.995

EXAMPLE 8

Association between smoking and lung cancer. This example is a further illustration of the use of OUTLIER for a set of contingency tables indexed by unordered categories. The data were originally given by Dorn (1954) and also analyzed by Cornfield (1956). Cox (1970) considers a model in which the logistic difference is the same for k independent 2 x 2 contingency tables. He defines a residual which should behave approximately like the residuals for a random sample from the unit normal distribution. He illustrates his graphical analysis with the same data as this example.

ASSOCIATION BETWEEN SMOKING AND LUNG CANCER

In Table 1a are listed the observations from 14 retrospective studies on the possible association between smoking and lung cancer. We denote the occurrences in the three-way 14 x 2 x 2 contingency table by x(ijk) with the notation

Characteristic	Index	1	2	3	...	14
Study	i	No. 1	No. 2	No. 3	...	No. 14
Patients	j	Control	Lung cancer			
Smoking	k	Nonsmoker	Smoker			

Does this data show association between smoking and lung cancer, and if so, is the association homogeneous over the 14 studies? Here the measure of association is the logarithm of the cross-product ratio.

TABLE 1a

Fourteen Retrospective Studies On The Association Between Smoking and Lung Cancer

	Control Patients		Lung Cancer Patients	
Study	Non-Smokers	Smokers	Non-Smokers	Smokers
1	14	72	3	83
2	43	227	3	90
3	19	81	7	129
4	125	397	12	70
5	131	299	32	412
6	114	666	8	597
7	12	174	5	88
8	61	1296	7	1350
9	27	106	3	60
10	81	534	18	459
11	54	246	4	724
12	56	462	19	499
13	636	1729	39	451
14	28	259	5	260

The hypothesis of conditional independence given the study

$$H_a: \frac{p(ijk)}{p(i..)} = \frac{p(ij.)}{p(i..)} \frac{p(i.k)}{p(i..)}$$

imposes the constraints on the MDI estimate $x_a^*(ijk)$ that

$$x_a^*(ij.) = x(ij.) \text{ and } x_a^*(i.k) = x(i.k).$$

In fact $x_a^*(ijk)$ may be explicitly represented by

$$x_a^*(ijk) = x(ij.)x(i.k)/x(i..).$$

Similarly the hypothesis of no second-order interaction

$$H_2\text{:}\quad p(ijk) = a(ij)b(ik)c(jk)$$

imposes the constraints on the MDI estimate $x_2^*(ijk)$ that

$$x_2^*(ij.) = x(ij.),\ x_2^*(i.k) = x(i.k),\ x_2^*(.jk) = x(.jk).$$

The MDI estimate $x_2^*(ijk)$ cannot be represented as an explicit product of the observed marginals.

The associated Analysis of Information table permits us to test the goodness-of-fit of the MDI estimates, also the effect of adding the marginal constraint x(.jk) to the marginal constraints x(ij.) and x(i.k), or the significance of the common natural parameter associated with x(.jk) for all 14 studies

$$\tau_{11}^{jk} = \ln \frac{x_2^*(i11)x_2^*(i22)}{x_2^*(i12)x_2^*(i21)},\ i = 1, 2, \ldots, 14.$$

We recall that the log-odds of control to lung cancer for the MDI estimates x_a^* and x_2^* are given by the representations

$$\ln \frac{x_a^*(i1k)}{x_a^*(i2k)} = \tau_1^j + \tau_{i1}^{ij}$$

$$\ln \frac{x_2^*(i1k)}{x_2^*(i2k)} = \tau_1^j + \tau_{i1}^{ij} + \tau_{11}^{jk}$$

where the values of the natural parameters (taus) depend of course on x_a^* and x_2^*.

TABLE 2

Analysis of Information

Component due to	Information	D.F.
H_a:x(ij.),x(i.k)	$2I(x:x_a^*) = 549.74$	14
H_2:x(ij.),x(i.k),x(.jk)	$2I(x_2^*:x_b^*) = 494.55$	1
	$2I(x:x_2^*) = 55.19$	13

The value of $2I(x:x_a^*)$ suggests that the null hypothesis of no association between smoking and lung cancer conditioned on the study is false. This conditional hypothesis allows the accumulation of information from different studies without imposing the requirement that the population characteristics of each study be similar. The rejection of this conditional independence hypothesis is of course expected. Moreover, $2I(x:x_2^*)$ is also significant. The degree of departure from no second-order interaction is dependent on the study. Is this dependence the result of a small subset of the studies which are substantially different from the remainder, or does the departure vary along a continuum?

The value of $2I(x_2^*:x_b^*)$ suggests that, computed as above, the value of $\tau_{11}^{jk} = 1.687$ is significantly different from zero. The value of $2I(x:x_2^*)$ suggests that we reject the null hypothesis of no second-order interaction, that is, the model with a common value of the association τ_{11}^{jk}, is not a good fit. The values of x_2^* are given in Table 1b. We now proceed to determine the outliers.

TABLE 1b

$x_2^*(ijk)$

Study	Control Patients Non-Smokers	Control Patients Smokers	Lung Cancer Patients Non-Smokers	Lung Cancer Patients Smokers
1	14.01	71.99	2.99	83.01
2	42.86	227.14	3.14	89.86
3	19.99	80.11	6.01	129.99
4	132.16	389.83	4.84	77.16
5	130.03	300.00	32.97	410.99
6	105.06	674.94	16.94	588.06
7	15.47	170.53	1.54	91.47
8	57.06	1299.93	10.94	1346.08
9	27.15	105.85	2.85	60.15
10	85.21	529.79	13.79	463.21
11	38.62	261.39	19.38	708.62
12	62.23	455.77	12.77	505.23
13	643.32	1721.66	31.69	458.33
14	27.84	259.16	5.17	259.84

Examination of the computer output for x_2^* using all 14 studies showed a largest OUTLIER value of 18.14 for the cell (11,2,1). A new MDI estimate fitting the marginals x(ij.), x(i.k), x(.jk) and omitting the cell (11,2,1) was obtained. In fact Study 11 was omitted because with the constraints for the new MDI estimate $x_b^*(11,j,k) = x(11,j,k)$. Since this estimate yielded

$$2I(x:x_b^*) = 28.40, \text{ 12 D.F.}$$

the deletion procedure was continued. We summarize the results in Table 3 and Analysis of Information Table 4.

TABLE 3

Fitting x(ij.), x(i.k), x(.jk) with sequential deletion of studies

Study Nos.	Largest OUTLIER Cell	Value	Information	D.F.
1-14	(11,2,1)	18.14	$2I(x:x_2^*) = 55.19$	13
1-10,12-14	(6,2,1)	7.89	$2I(x:x_b^*) = 28.40$	12
1-5,7-10,12-14	(4,2,1)	4.87	$2I(x:x_c^*) = 18.03$	11
1-3,5,7-10,12-14	(7,2,1)	3.91	$2I(x:x_d^*) = 11.94$	10
1-3,5,8-10,12-14			$2I(x:x_e^*) = 7.03$	9

TABLE 4

Analysis of Information

Component due to	Information	D.F.
All 14 studies	$2I(x:x_2^*) = 55.19$	13
Less 11	$2I(x_b^*:x_2^*) = 26.79$	1
	$2I(x:x_b^*) = 28.40$	12
Less 11,6	$2I(x_c^*:x_b^*) = 10.37$	1
	$2I(x:x_c^*) = 18.03$	11
Less 11,6,4	$2I(x_d^*:x_c^*) = 6.09$	1
	$2I(x:x_d^*) = 11.94$	10
Less 11,6,4,7	$2I(x_e^*:x_d^*) = 4.91$	1
	$2I(x:x_e^*) = 7.03$	9

Since $(2I(x:x_2^*) - 2I(x:x_e^*))/2I(x:x_2^*) = 2I(x_e^*:x_2^*)/2I(x:x_2^*) = 48.16/55.19 = 0.87$ we see that the four studies numbered 4,6,7,11 contributed 87% of the unexplained variation $2I(x:x_2^*)$. The values of the MDI estimate x_e^* are given in Table 1c. The value of the log cross-product ratio is

$$\tau_{11}^{jk} = \ell n \frac{x_e^*(i11)x_e^*(i22)}{x_e^*(i12)x_e^*(i21)} = 1.55,\ i = 1\text{-}3,5,8\text{-}10,12\text{-}14.$$

TABLE 1c

$x_e^*(ijk)$

	Control Patients		Lung Cancer Patients	
Study	Non-Smokers	Smokers	Non-Smokers	Smokers
1	13.69	72.32	3.32	82.69
2	42.46	227.53	3.54	89.47
3	19.40	80.60	6.60	129.40
5	126.85	303.18	36.15	407.83
8	55.79	1301.21	12.21	1344.80
9	26.80	106.20	3.20	59.80
10	83.61	531.39	15.39	461.62
12	60.81	457.20	14.19	503.81
13	639.35	1725.64	35.66	454.36
14	27.24	259.76	5.76	259.24

We note that Cox (1970) in analyzing the data of Table 1a concluded that Studies 8, 6, and 11 were outliers. For the 14 studies he found a residual sum of squares 47.7 with 13 degrees of freedom. With Studies 8, 6, and 11 omitted he found a residual sum of squares 15.1 with 10 degrees of freedom. (Cox (1970) p. 83 gives the degrees of freedom as 11, a misprint.)

Following the procedure as above, when Studies 6, 8, and 11 were omitted the results led to the analysis of information Table 5.

TABLE 5

Analysis of Information

Component due to	Information	D.F.
All 14 studies	$2I(x:x_2^*) = 55.19$	13
Less 6,8,11	$2I(x_f^*:x_2^*) = 41.62$	3
	$2I(x:x_f^*) = 13.57$	10

Comparison with Table 4 shows that omitting Studies 6, 4, and 11 yields better values than omitting Studies 6, 8, and 11.

The sequential procedure discussed herein was also applied to data relating father and son professions published by Karl Pearson (1904) and considered by Fienberg (1969) and Good (1956). Using the Pearson data Fienberg obtained an $X^2 = 184.9$ with 44 out of 196 cells deleted whereas the sequential procedure led to an $X^2 = 155.3$ with 25 cells deleted.

EXAMPLE 9

Zero marginals. This example illustrates what can happen when zero marginals are included in the fitting procedure. Also discussed is the necessary modification in the degrees of freedom. The procedure is related to that discussed with respect to OUTLIERS.

ZERO MARGINALS

As may be noted from the earlier analyses, zero occurrences in cells of the observed contingency table present no special problem in ICP provided that no marginal entering into the fitting constraints is zero. When such is the case, however, the interpretation may be distorted because of inflated degrees of freedom. A procedure to circumvent this problem is similar to that used for getting revised MDI estimates when outliers are indicated. We shall present the procedure in terms of a specific example.

The following data resulted from a study of Christmas tree consumption. We are indebted to Dipl. Forstwirt Dietrich v. Staden, Institut f. Forstbenutzung, Universitaet Goettingen for the data and permission to use it. In Table 1 are listed responses to the question "Did you have a Christmas tree in your apartment/house last year or not?", according to size of household and size of city. We denote the occurrences in the three-way 2 x 9 x 5 contingency table by x(ijk) with the notation

Characteristic	Index	1	2	3	4	5	6	7	8	9
Tree	i	Yes	No							
Household size	j	1	2	3	4	5	6	7	8	9
City size	k	< 2000	2000 to 20000	20000 to 100000	100000 to 500000	500000 or more				

For a 2 x 9 x 5 R x C x D contingency table we compute an MDI estimate under a hypothesis of no second-order interaction by fitting all the two-way marginals. Call this MDI estimate $x_2^*(ijk)$. A test for the null hypothesis of no second-order interaction is given by

$$2\Sigma\Sigma\Sigma\ x(ijk)\ \ln \frac{x(ijk)}{x_2^*(ijk)} = 2I(x:x_2^*),\ 32\ \text{D.F.}$$

If there is no second-order interaction then the associations between R and C, R and D, C and D are the same for all values of the third variable, that is,

$$\ln \frac{x_2^*(1jk)x_2^*(29k)}{x_2^*(2jk)x_2^*(19k)} \text{ depends only on } j,$$

$$\ln \frac{x_2^*(1jk)x_2^*(2j5)}{x_2^*(2jk)x_2^*(1j5)} \text{ depends only on } k,$$

$$\ln \frac{x_2^*(ijk)x_2^*(i95)}{x_2^*(ij5)x_2^*(i9k)} \text{ is independent of } i.$$

Within this model a test whether the marginal x(i.k) contributes significantly is obtained by computing an MDI estimate fitting the marginals x(ij.), x(.jk) only. Call this MDI estimate $x_b^*(ijk)$, which can be expressed explicitly as $x_b^*(ijk) = x(ij.)x(.jk)/x(.j.)$. We recognize $x_b^*(ijk)$ as the MDI estimate under an hypothesis of conditional independence of R and D given C. We now have Analysis of Information Table 2.

TABLE 2

Component due to	Information	D.F.
Conditional independence of R and D given C	$2I(x:x_b^*)$	36
Effect of x(i.k) given x(ij.) and x(.jk)	$2I(x_2^*:x_b^*)$	4
No second-order interaction	$2I(x:x_2^*)$	32

The MDI estimate $x_2^*(ijk)$ satisfies the constraints

$$x_2^*(1jk) + x_2^*(2jk) = x(1jk) + x(2jk)$$

$$x_2^*(1j1) + x_2^*(1j2) + \ldots + x_2^*(1j5) = x(1j1) + x(1j2) + \ldots + x(1j5)$$

$$x_2^*(2j1) + x_2^*(2j2) + \ldots + x_2^*(2j5) = x(2j1) + x(2j2) + \ldots + x(2j5).$$

Since x(2jk) = 0 for j = 6, 7, 8, 9, it follows that

$$x_2^*(2j1) + \ldots + x_2^*(2j5) = 0 \text{ for } j = 6,7,8,9.$$

But $x_2^*(ijk) \geqq 0$, hence $x_2^*(2jk) = 0$ for j = 6,7,8,9, and $x_b^*(1jk) = x(1jk)$ for j = 6,7,8,9. Accordingly let us compute an MDI estimate $x_f^*(ijk)$ by fitting the two-way marginals of the 2 x 5 x 5 table j = 1,2,3,4,5 and $x_f^*(ijk) = x(ijk)$, j = 6,7,8,9. Similarly get the MDI estimate $x_e^*(ijk) = x(ij.)x(.jk)/x(.j.)$ for the 2 x 5 x 5 table j = 1,2,3,4,5 and $x_e^*(ijk) = x(ijk)$, j = 6,7,8,9.

We now find

TABLE 3

Component due to	Information	D.F.
Conditional independence of R and D given C	$2I(x:x_e^*) = 25.532$	20
Effect of x(i.k) given x(ij.) and x(.jk)	$2I(x_f^*:x_e^*) = 5.821$	4
No second-order interaction	$2I(x:x_f^*) = 19.711$	16

Note the reduction in the degrees of freedom between Table 2 and Table 3. It is also interesting to note that the MDI estimation procedures for Table 2 yielded the same MDI estimates and statistics as for Table 3. See Tables 4 and 5, Tables 6 and 7.

Since the MDI estimate x_e^* provides a good fit to the data, it seems reasonable to conclude that the purchase of a Christmas tree is independent of the size of the city given the size of the household (j = 1,2,3,4,5) and households of size 6,7,8,9 seem almost sure to buy Christmas trees.

The log-odds and the odds for the purchase of a Christmas tree as a function of household size are given in Table 8.

TABLE 8

$\ln(x_e^*(1jk)/x_e^*(2jk)) =$ $\ln(x(1j.)/x(2j.))$		$x_e^*(1jk)/x_e^*(2jk) =$ $x(1j.)/x(2j.)$	
j		j	
1	-0.2586	1	0.772
2	0.8662	2	2.378
3	2.1702	3	8.760
4	3.4012	4	30.000
5	2.3716	5	10.715

TABLE 1

Original Data x(ijk)

ij	k = 1	2	3	4	5	
11	4	7	6	12	32	61
12	18	37	55	41	63	214
13	20	45	52	41	61	219
14	25	40	38	32	45	180
15	11	31	14	11	8	75
16	4	8	12	1	6	31
17	3	0	2	1	1	7
18	2	0	1	0	1	4
19	0	1	0	0	1	2
21	4	13	9	13	40	79
22	5	18	19	15	33	90
23	3	8	0	4	10	25
24	0	3	1	0	2	6
25	1	1	2	2	1	7
26	0	0	0	0	0	0
27	0	0	0	0	0	0
28	0	0	0	0	0	0
29	0	0	0	0	0	0

TABLE 4

$x_2^*(ijk)$ 2 x 9 x 5

k = 1	2	3	4	5
4.066	7.835	7.815	12.046	29.240
17.252	35.836	56.204	40.865	63.844
21.123	46.382	47.951	40.955	62.588
24.364	41.268	38.056	31.098	45.213
11.193	28.680	14.974	12.036	8.115
4.000	8.000	12.000	1.000	6.000
3.000	0.000	2.000	1.000	1.000
2.000	0.000	1.000	0.000	1.000
0.000	1.000	0.000	0.000	1.000
3.933	12.167	7.184	12.953	42.763
5.747	19.166	17.795	15.134	32.157
1.877	6.618	4.050	4.046	8.410
0.636	1.731	0.945	0.903	1.786
0.807	3.319	1.026	0.964	0.884
0.000	0.000	0.000	0.000	0.000
0.000	0.000	0.000	0.000	0.000
0.000	0.000	0.000	0.000	0.000
0.000	0.000	0.000	0.000	0.000

TABLE 5

$x_f^*(ijk)$

ij	k = 1	2	3	4	5
11	4.066	7.835	7.815	12.046	29.240
12	17.253	35.835	56.204	40.865	63.843
13	21.123	46.382	47.915	40.955	62.589
14	24.364	41.268	38.056	31.098	45.213
15	11.193	28.680	14.974	12.036	8.115
16	4	8	12	1	6
17	3	0	2	1	1
18	2	0	1	0	1
19	0	1	0	0	1
21	3.933	12.167	7.184	12.953	42.763
22	5.747	19.166	17.795	15.134	32.157
23	1.877	6.618	4.050	4.046	8.410
24	0.636	1.731	0.945	0.903	1.786
25	0.807	3.319	1.026	0.964	0.884
26	0	0	0	0	0
27	0	0	0	0	0
28	0	0	0	0	0
29	0	0	0	0	0

TABLE 6

$x_b^*(ijk) = x(ij.)x(.jk)/x(.j.)$ 2 x 9 x 5

	k				
ij	1	2	3	4	5
11	3.486	8.714	6.536	10.893	31.371
12	16.191	38.717	52.092	39.421	67.579
13	20.643	47.570	46.672	40.389	63.725
14	24.194	41.613	37.742	30.968	45.484
15	10.976	29.268	14.634	11.890	8.232
16	4.000	8.000	12.000	1.000	6.000
17	3.000	0.000	2.000	1.000	1.000
18	2.000	0.000	1.000	0.000	1.000
19	0.000	1.000	0.000	0.000	1.000
21	4.514	11.286	8.464	14.107	40.629
22	6.809	16.283	21.908	16.579	28.421
23	2.357	5.430	5.328	4.611	7.275
24	0.806	1.387	1.258	1.032	1.516
25	1.024	2.732	1.366	1.110	0.768
26	0.000	0.000	0.000	0.000	0.000
27	0.000	0.000	0.000	0.000	0.000
28	0.000	0.000	0.000	0.000	0.000
29	0.000	0.000	0.000	0.000	0.000

TABLE 7

$x_e^*(ijk)$

	k				
ij	1	2	3	4	5
11	3.486	8.714	6.536	10.893	31.371
12	16.191	38.717	52.092	39.421	67.579
13	20.643	47.570	46.672	40.389	63.725
14	24.194	41.613	37.742	30.968	45.484
15	10.976	29.268	14.634	11.890	8.232
16	4	8	12	1	6
17	3	0	2	1	1
18	2	0	1	0	1
19	0	1	0	0	1
21	4.514	11.286	8.464	14.107	40.629
22	6.809	16.283	21.908	16.579	28.421
23	2.357	5.430	5.328	4.611	7.275
24	0.806	1.387	1.258	1.032	1.516
25	1.024	2.732	1.366	1.110	0.768
26	0	0	0	0	0
27	0	0	0	0	0
28	0	0	0	0	0
29	0	0	0	0	0

Chapter 5

ANALYSIS: GENERAL FORM

I. GENERAL LINEAR HYPOTHESIS

In Chapter 3, the loglinear representation obtained by applying the principle of MDI estimation was examined with particular emphasis on problems of fitting or smoothing contingency tables using a set of observed marginals and/or linear combinations of observed cell entries as fitting constraints. The matrix $\tilde{C}$ and vector $\tilde{\theta}$ of the moment constraints $\tilde{C}\tilde{p} = \tilde{\theta}$ were specialized in Section IIB equation (8) of Chapter 3 as $\tilde{T}'\tilde{p}^* = \tilde{T}'\tilde{\pi}_0$ or $\tilde{T}'\tilde{x}^* = \tilde{T}'\tilde{x}$ where $\tilde{x} = n\tilde{\pi}_0$ is the observed distribution. Further, for marginal constraints the $T(\omega)$ functions were indicator functions and hence took values 0 and 1 only.

We now propose to consider problems of a general and different nature. In these problems the moment constraints are determined by external hypotheses of interest and one is concerned whether <u>the observed distribution is consistent therewith</u>. The observed distribution, obviously, does not satisfy the moment constraints (if it does it will be consistent, *ipso facto*) and the problem is to determine whether its departure from the moment constraints can be attributed to chance variations. The objective is then to find an MDI estimate that satisfies the given moment constraints and is as similar as possible to the observed distribution, that is, with discrimination information between the two distributions minimized. The MDI estimate then conserves all the properties of the observed distribution except those implied by the given moment constraints. Such problems are called external constraints problems (ECP). External constraints problems can be divided into two cases; one in which the observed distribution satisfies none of the given moment constraints and the other in which it satisfies some (but not all) of the given moment constraints. In the latter case, for example, the MDI estimate may be constrained to have some set of marginals the same as the observed distribution, but have some other set of marginals or linear combination of cell values satisfy arbitrary values based on external considerations.

Before presenting the general analysis in symbolic form, we give in the next section several examples of ECP. Section III presents the general analysis of the single sample case. In Section IV this method of analysis is fully illustrated with the help of numerical data of an example from Section II. Sections V and VI deal with the analysis in the k-sample case and an illustration. In Section VII a related method of estimation, minimum modified chi-square is discussed.

For the k-sample problems the moment constraints are denoted as $\tilde{B}\tilde{p} = \tilde{\theta}$ rather than as $\tilde{C}\tilde{p} = \tilde{\theta}$ to retain consistency since the B-matrix will be suitably transformed so that methods of analysis of the single sample case are applicable. The transformed matrix is then denoted by $\tilde{C}$, as in the one sample case. The purpose of Section II is to explain how to arrive at the moment constraints expressed in the form $\tilde{B}\tilde{p} = \tilde{\theta}$.

II. EXAMPLES: EXTERNAL CONSTRAINTS PROBLEM (ECP)

In this section, we give several simple examples in which MDI analysis is applicable. For count data with quantitative values of the random variable, we consider the following hypotheses for several samples:

(i) equality of specified population means

(ii) equality of population means

(iii) equality of population means <u>and</u> variances.

For count data represented as a set of contingency tables, the random variables need not be necessarily quantitative (e.g. eye-color). But even in such categorical cases, the MDI approach can be used to analyze the data under many hypotheses, including those of

(i) specified marginals,

(ii) homogeneity of certain multinomials (marginals)

(iii) equality of marginal probabilities and/or individual probabilities in several contingency tables,

(iv) no interaction on a linear scale, that is, cell probabilities can be expressed as <u>linear</u> functions of parameters which are structurally less complex.

An important step in the analysis of all such problems is the formulation of the hypothesis in terms of linear constraints on the underlying probabilities. This is illustrated below. For the sake of brevity, the three hypotheses relating to count data on quantitative random variables are discussed with the help of one example. In multidimensional contingency tables, with the same number of categories for each dimension, MDI analysis in ECP is applicable under hypotheses of symmetry and marginal homogeneity. For a different formulation of such problems see Ireland, Ku, and Kullback (1969) and Kullback (1971a, 1971b).

<u>Ex. 1</u>: In animal-trapping experiments, a certain standard type of trap can hold at most three animals. A new trap, which can hold up to five animals, is under trial. Two independent random samples of respective sizes 100 and 120 traps are laid out under homogeneous conditions. The results (artificial) are shown below:

TABLE 1

Animal trapping data on two trap-types

No. of animals trapped		0	1	2	3	4	5	Total
No. of traps:	Standard	38	34	20	8	-	-	100
No. of traps:	New	37	32	31	9	8	3	120

Consider the following hypotheses:

H_1: The population means for both the traps are equal to 1.

H_2: The two population means are equal.

H_3: The two populations have the same means and variances.

There are two independent multinomial experiments, the first with four cells and the second with six. Let us denote the random variables (number of animals trapped) in the first experiment by $y(1i)$, $i = 1, 2, 3, 4$, thus $y(11) = 0$, $y(12) = 1$, $y(13) = 2$, $y(14) = 3$. The corresponding cell probabilities are denoted by $p(1i)$, $i = 1, 2, 3, 4$. In the second experiment with the new trap, the random variable (number of animals trapped) are denoted by $y(2i)$, $i = 1, \ldots, 6$ thus $y(21) = 0$, $y(22) = 1$, $y(23) = 2$, $y(24) = 3$, $y(25) = 4$, $y(26) = 5$. The corresponding cell probabilities are denoted by $p(2i)$, $i = 1, \ldots, 6$. Note that the random variables are quantitative.

There are two <u>natural</u> constraints on the probabilities: $\sum_i p(1i) = 1$ and $\sum_i p(2i) = 1$. The hypothesis H_1 that the population means are both equal to 1 implies further linear constraints given by

$$\Sigma_i \; y(1i)p(1i) = 1 \tag{1}$$

$$\Sigma_i \; y(2i)p(2i) = 1. \tag{2}$$

It is convenient to express the constraints in matrix notation. For this purpose, let

$$\tilde{p}' = (p(11), p(12), p(13), p(14), p(21), p(22), p(23), p(24), p(25), p(26)). \tag{3}$$

Then the constraints can be expressed in the form $\underset{\sim}{B}_1\underset{\sim}{p} = \underset{\sim}{\theta}_1$, say where

$$\underset{\sim}{B}_1 = \begin{bmatrix} 1 & 1 & 1 & 1 & 0 & 0 & 0 & 0 & 0 & 0 \\ 0 & 0 & 0 & 0 & 1 & 1 & 1 & 1 & 1 & 1 \\ 0 & 1 & 2 & 3 & 0 & 0 & 0 & 0 & 0 & 0 \\ 0 & 0 & 0 & 0 & 0 & 1 & 2 & 3 & 4 & 5 \end{bmatrix}$$

and

$$\underset{\sim}{\theta}_1' = (1,1,1,1).$$

The first two rows of $\underset{\sim}{B}_1$ express the natural constraints and the third and fourth row correspond to (1) and (2).

In hypothesis H_2, the common value of the population mean is not specified. Hence H_2 imposes only one constraint in addition to the natural ones. It is

(4) $\Sigma_i\ y(1i)p(1i) = \Sigma_i\ y(2i)p(2i)$ i.e. $\Sigma_i\ y(1i)p(1i) -$

$$\Sigma_i\ y(2i)p(2i) = 0.$$

The matrices $\underset{\sim}{B}_2$ and $\underset{\sim}{\theta}_2$ corresponding to H_2 can be written as

$$\underset{\sim}{B}_2 = \begin{bmatrix} 1 & 1 & 1 & 1 & 0 & 0 & 0 & 0 & 0 & 0 \\ 0 & 0 & 0 & 0 & 1 & 1 & 1 & 1 & 1 & 1 \\ 0 & 1 & 2 & 3 & 0 & -1 & -2 & -3 & -4 & -5 \end{bmatrix}$$

and

$$\theta_2' = (1,\ 1,\ 0).$$

Hypothesis H_3 postulates equality of means and variances. Given that the means are equal, equality of variances is the same as equality of second moments about the origin. Using this fact, we can write the linear constraints corresponding to H_3 as

(5) $$\Sigma_i\ y(1i)p(1i) - \Sigma_i\ y(2i)p(2i) = 0$$

(6) $$\Sigma_i\ y^2(1i)p(1i) - \Sigma_i\ y^2(2i)p(2i) = 0.$$

Thus

$$B_3 = \begin{bmatrix} 1 & 1 & 1 & 1 & 0 & 0 & 0 & 0 & 0 & 0 \\ 0 & 0 & 0 & 0 & 1 & 1 & 1 & 1 & 1 & 1 \\ 0 & 1 & 2 & 3 & 0 & -1 & -2 & -3 & -4 & -5 \\ 0 & 1 & 4 & 9 & 0 & -1 & -4 & -9 & -16 & -25 \end{bmatrix}$$

and

$$\theta_3' = (1, 1, 0, 0).$$

Note that the hypothesis H_1 could also have been stated in terms of the constraints (1) and (4). This shows that there may be several equivalent formulations of the hypothesis under study. Of course, the estimates of cell frequencies and the test statistics are unchanged under different but equivalent formulations of the hypotheses. The loglinear representation of the estimates however involves parameters whose values do depend on the formulation. Note that H_1 can also be stated in terms of constraints (1), (2), and (4). But the three are not linearly independent. Any two imply the remaining one. By convention, we specify H_1 in terms of linearly independent constraints only.

It may be pointed out that H_1 implies H_2. Hypothesis H_3 also implies H_2. Thus H_2 is a weaker hypothesis than H_1 or H_3. This fact is interestingly reflected in their corresponding B-matrices. If H_1 is formulated in terms of constraints (1) and (4), the rows of its B-matrix contain all the rows of B_2. Similarly the rows of B_3 contain all the rows of B_2. If an hypothesis implies another (weaker) hypothesis, we can always find two B-matrices corresponding to the two hypotheses such that all the rows of the B-matrix of the weaker hypothesis are contained in the B-matrix of the other.

Remarks in the above two paragraphs apply to contingency tables also. If we construct a C-matrix consisting of the first four columns of the first and third rows of B_1 and let $\theta' = (1,1)$ the constraints $Cp = \theta$ correspond to a single sample hypothesis that the population mean corresponding to the standard trap is 1.

Ex. 2: In the data given by Fisher (1954, 244), the following 2 x 2 table represents seedling counts on self-fertilized heterozygotes for two factors in maize, starchy vs. sugary and green vs. white base leaf.

	Green	White
Starchy	1997	906
Sugary	904	32

In accordance with genetic theory, the marginal probabilities for Starchy category and for Green category should be 0.75 each. The cells of the contingency table are indexed in lexicographic order as (11, 12, 21, 22).

We now have a single multinomial experiment with four cells. The values of the random variable are qualitative; (Starchy, Green), (Starchy, White), (Sugary, Green) and (Sugary, White). But these value do not appear in the constraints so the approach is still applicable. The probabilities $p(ij)$, $i = 1, 2$, $j = 1, 2$, obey the following constraints specified by the genetic hypothesis

$$p(11) + p(12) = 0.75 \tag{7}$$

$$p(11) + p(21) = 0.75. \tag{8}$$

With $\underset{\sim}{p}' = (p(11), p(12), p(21), p(22))$, the constraints (7) and (8) together with the natural constraint $\Sigma\,\Sigma\, p(ij) = 1$ are put in the form $\underset{\sim}{C}\underset{\sim}{p} = \underset{\sim}{\theta}$ by letting

$$\underset{\sim}{C} = \begin{bmatrix} 1 & 1 & 1 & 1 \\ 1 & 1 & 0 & 0 \\ 1 & 0 & 1 & 0 \end{bmatrix}$$

and

$$\underset{\sim}{\theta}' = (1, 0.75, 0.75).$$

The constraint $p(21) + p(22) = 0.25$, for example, is not included since it is implied by the natural constraint and (7). This data was also considered in Ireland and Kullback (1968b) using the Deming-Stephan iterative scaling algorithm.

<u>Ex. 3</u>: Homogeneity of binomials: The following table is based on the data of the hourly distribution of live and still births from Hill (1955).

Time Interval	Number of live births	Number of still births	Total
Midnight -	8691	283	8974
6 A.M. -	8839	337	9176
12 Noon -	6892	328	7220
6 P.M. -	7802	324	8126

For the sake of illustration the data are regarded as independent samples from 4 binomial populations corresponding to the 4 time intervals of the day. The hypothesis of interest is that probabilities of live births are the same for the four intervals. Writing the probabilities in lexicographic order,

$$\underset{\sim}{p}' = (p(11), p(12), p(21), p(22), p(31), p(32), p(41), p(42))$$

the hypothesis can be expressed as

$$p(11) = p(21) = p(31) = p(41). \tag{9}$$

The equalities in (9) represent three linearly independent constraints which can be taken to be

$$p(11) = p(21) \tag{10}$$

$$p(11) = p(31) \tag{11}$$

$$p(11) = p(41). \tag{12}$$

For this hypothesis, the formulation $\underset{\sim}{B}\underset{\sim}{p} = \underset{\sim}{\theta}$ is achieved by setting

$$\underset{\sim}{B} = \begin{bmatrix} 1 & 1 & 0 & 0 & 0 & 0 & 0 & 0 \\ 0 & 0 & 1 & 1 & 0 & 0 & 0 & 0 \\ 0 & 0 & 0 & 0 & 1 & 1 & 0 & 0 \\ 0 & 0 & 0 & 0 & 0 & 0 & 1 & 1 \\ 1 & 0 & -1 & 0 & 0 & 0 & 0 & 0 \\ 1 & 0 & 0 & 0 & -1 & 0 & 0 & 0 \\ 1 & 0 & 0 & 0 & 0 & 0 & -1 & 0 \end{bmatrix}$$

and

$$\underset{\sim}{\theta}' = (1, 1, 1, 1, 0, 0, 0).$$

Ex. 4: No interaction on a linear scale: (See Chapter 7.) In the data given by Bartlett (1935) are found the results of an experiment designed to investigate the propagation of plum root stocks from root cuttings. There were 240 cuttings for each of the four treatments.

	At Once $i = 1$		In Spring $i = 2$	
	Long $j = 1$	Short $j = 2$	Long $j = 1$	Short $j = 2$
Dead $k = 1$	84	133	156	209
Alive $k = 2$	156	107	84	31
	240	240	240	240

Thus there are four independent binomials. The null hypothesis of no interaction on a linear scale is that

$$p(ijk) = \mu(..k) + \mu(i.k) + \mu(.jk)$$

or

$$H_0: \ p(111) - p(121) - p(211) + p(221) = 0,$$

which can be expressed in the $\underset{\sim}{B}\underset{\sim}{p} = \underset{\sim}{\theta}$ formulation by setting

$$\underset{\sim}{p}' = (p(111), p(112), p(121), p(122), p(211), p(212), p(221), p(222)),$$

$$\underset{\sim}{B} = \begin{bmatrix} 1 & 1 & 0 & 0 & 0 & 0 & 0 & 0 \\ 0 & 0 & 1 & 1 & 0 & 0 & 0 & 0 \\ 0 & 0 & 0 & 0 & 1 & 1 & 0 & 0 \\ 0 & 0 & 0 & 0 & 0 & 0 & 1 & 1 \\ 1 & 0 & -1 & 0 & -1 & 0 & 1 & 0 \end{bmatrix}$$

and

$$\underset{\sim}{\theta}' = (1, 1, 1, 1, 0).$$

This problem was also considered by Snedecor and Cochran (1967), Bhapkar and Koch (1968), Berkson (1972), Kullback (1974).

Ex. 5: Several contingency tables, equality of marginal and individual probabilities: In the data given by Gail (1974, 97) there are two 2 x 3 contingency tables given by

$$\begin{matrix} 20 & 5 & 3 \\ 6 & 4 & 2 \end{matrix} \qquad \begin{matrix} 15 & 15 & 2 \\ 10 & 5 & 5 \end{matrix}.$$

Thus there are two independent multinomial experiments, each with six cells.

The hypothesis H_1 of equality of marginal probabilities is expressed in terms of probabilities $p(hij)$, $h = 1, 2$, $i = 1, 2$, $j = 1, 2, 3$, as

$$p(111) + p(112) + p(113) = p(211) + p(212) + p(213),$$

$$p(111) + p(121) = p(211) + p(221),$$

$$p(112) + p(122) = p(212) + p(222),$$

which can be expressed in the $\underset{\sim}{B}\underset{\sim}{p} = \underset{\sim}{\theta}$ formulation by letting

$$\underset{\sim}{p}' = (p(111), p(112), p(113), p(121), p(122), p(123), p(211), p(212), p(213), p(221), p(222), p(223)),$$

$$\underset{\sim}{B} = \begin{bmatrix} 1 & 1 & 1 & 1 & 1 & 1 & 0 & 0 & 0 & 0 & 0 & 0 \\ 0 & 0 & 0 & 0 & 0 & 0 & 1 & 1 & 1 & 1 & 1 & 1 \\ 1 & 1 & 1 & 0 & 0 & 0 & -1 & -1 & -1 & 0 & 0 & 0 \\ 1 & 0 & 0 & 1 & 0 & 0 & -1 & 0 & 0 & -1 & 0 & 0 \\ 0 & 1 & 0 & 0 & 1 & 0 & 0 & -1 & 0 & 0 & -1 & 0 \end{bmatrix}$$

and

$$\underset{\sim}{\theta}' = (1, 1, 0, 0, 0).$$

The hypothesis H_2 of equality of individual probabilities specified by the constraints

$$p(1ij) = p(2ij),\ i = 1, 2,\ j = 1, 2, 3,$$

can be expressed as $\underset{\sim}{B}\underset{\sim}{p} = \underset{\sim}{\theta}$ with

$$\underset{\sim}{B} = \begin{bmatrix} 1 & 1 & 1 & 1 & 1 & 1 & 0 & 0 & 0 & 0 & 0 & 0 \\ 0 & 0 & 0 & 0 & 0 & 0 & 1 & 1 & 1 & 1 & 1 & 1 \\ 1 & 0 & 0 & 0 & 0 & 0 & -1 & 0 & 0 & 0 & 0 & 0 \\ 0 & 1 & 0 & 0 & 0 & 0 & 0 & -1 & 0 & 0 & 0 & 0 \\ 0 & 0 & 1 & 0 & 0 & 0 & 0 & 0 & -1 & 0 & 0 & 0 \\ 0 & 0 & 0 & 1 & 0 & 0 & 0 & 0 & 0 & -1 & 0 & 0 \\ 0 & 0 & 0 & 0 & 1 & 0 & 0 & 0 & 0 & 0 & -1 & 0 \end{bmatrix}$$

and

$$\underset{\sim}{\theta}' = (1, 1, 0, 0, 0, 0, 0)$$

using constraints which are linearly independent. See Chap. 8, Ex. 1.

III. GENERAL FORMULATION AND MDI ANALYSIS (SINGLE SAMPLE CASE)

For convenience, we restate the general problem. The discrimination information between two distributions $p(\omega)$ and $\pi(\omega)$, $\omega \in \Omega$, given by

$$I(p:\pi) = \sum_{\Omega} p(\omega) \ln (p(\omega)/\pi(\omega)) \tag{13}$$

is to be minimized over the family of distributions which satisfy the linearly independent moment constraints, for $\pi(\omega)$ fixed,

$$\underset{\sim}{C}\underset{\sim}{p} = \underset{\sim}{\theta} \tag{14}$$

where the design matrix $\underset{\sim}{C}$ is $(r + 1) \times \Omega$, $\underset{\sim}{p}$ is $\Omega \times 1$, $\underset{\sim}{\theta}$ is $(r + 1) \times 1$ and the rank of $\underset{\sim}{C}$ is $r + 1 \leq \Omega$. Both $\underset{\sim}{C}$ and $\underset{\sim}{\theta}$ are assumed to be known.

Differentiation of (13) with respect to $p(\omega)$, subject to (14), using undetermined Lagrange multipliers yields the minimizing distribution

$$p^*(\omega) = \exp(\tau_0 c_0(\omega) + \tau_1 c_1(\omega) + \dots + \tau_r c_r(\omega))\pi(\omega) \tag{15}$$

or

$$\ln (p^*(\omega)/\pi(\omega)) = \tau_0 c_0(\omega) + \tau_1 c_1(\omega) + \dots + \tau_r c_r(\omega), \quad \omega = 1, \dots, \Omega. \tag{16}$$

We can write (16) in matrix notation as

$$(17) \qquad \ln(p^*/\pi) = C'\tau.$$

The r + 1 taus, Lagrangian multipliers, are the natural or exponential parameters of the model, and are to be determined so that p^* satisfies $Cp^* = \theta$.

We consider the partitioning of the matrix C in (14) as follows:

$$C = \begin{bmatrix} C_1 \\ C_2 \end{bmatrix} \text{where } C_1 \text{ is } 1 \times \Omega,\ C_2 \text{ is } r \times \Omega,$$

$$\theta = \begin{bmatrix} 1 \\ \theta^* \end{bmatrix} \text{where } \theta^* \text{ is } r \times 1,$$

that is $C_1p = 1$, $C_2p = \theta^*$.

In the ECP, the observed distribution does not satisfy the moment constraints $Cp = \theta$. But rather than fitting or smoothing the observed data as in the case of the ICP, the objective now is to determine from amongst all the distributions p which satisfy $Cp = \theta$, the one which is closest to the observed distribution in the MDI sense. We take $\pi(\omega) = x(\omega)/N$, where $N = \sum_\Omega x(\omega)$. Setting $x^*(\omega) = Np^*(\omega)$, the MDI statistic for ECP is

$$(18) \qquad 2I(x^*:x) = 2\sum_\Omega x^*(\omega)\ln(x^*(\omega)/x(\omega))$$

which is asymptotically distributed as a central chi-square with r degrees of freedom if the observed table $x(\omega)$ is consistent with the null hypothesis (14) or the estimate $x^*(\omega)$.

In accordance with the discussion in Chapter 3 section VIII C, a quadratic approximation to $2I(x^*:x)$ is

$$(19) \qquad 2I(x^*:x) \simeq (N\theta^* - N\hat{\theta})' S_{22.1}^{-1}(N\theta^* - N\hat{\theta}),$$

where $C_1\pi = 1$, $C_2\pi = \hat{\theta}$, $S = \begin{bmatrix} C_1 \\ C_2 \end{bmatrix} D_x(C_1', C_2')$

$$= \begin{bmatrix} C_1D_xC_1' & C_1D_xC_2' \\ C_2D_xC_1' & C_2D_xC_2' \end{bmatrix} = \begin{bmatrix} S_{11} & S_{12} \\ S_{21} & S_{22} \end{bmatrix},\ S_{22.1} = S_{22} - S_{21}S_{11}^{-1}S_{12},$$

and $D_{\tilde{x}}$ is the $\Omega \times \Omega$ diagonal matrix with main diagonal entries $x(\omega)$ in lexicographic order. We shall see in Section VII that the righthand side of (19) is the minimum modified chi-square.

A. Analysis Of Information

For the ICP, we described how the MDI statistic corresponding to the comparison of an observed contingency table with an MDI estimate can be partitioned into two components; one representing a measure of effect, i.e. the effect of fitting additional marginals (or their associated natural parameters) and a measure of interaction or goodness-of-fit that represents the MDI statistic corresponding to the MDI estimate in which the enlarged set of moment constraints is used. (See Chapter 3 for details.) A similar analysis of information can be employed for the ECP.

Suppose that two hypotheses H_1 and H_2 are being considered in a problem. Let H_1 correspond to $\tilde{B}_1\tilde{p} = \tilde{\theta}_1$ and H_2 to $\tilde{B}_2\tilde{p} = \tilde{\theta}_2$. Suppose further that H_2 is stronger than H_1, that is, the moment constraints $\tilde{B}_2\tilde{p} = \tilde{\theta}_2$ imply $\tilde{B}_1\tilde{p} = \tilde{\theta}_1$. If m_2 is the number of external constraints in $\tilde{B}_2$ and m_1, those in $\tilde{B}_1$, then $m_2 > m_1$. If $2I(x_1^*:x)$ and $2I(x_2^*:x)$ are the MDI statistics corresponding to H_1 and H_2, they are respectively asymptotically distributed as chi-square with m_1 and m_2 degrees of freedom. Further

$$2I(x_2^*:x) = 2I(x_2^*:x_1^*) + 2I(x_1^*:x)$$

and $2I(x_2^*:x_1^*)$, which measures the contribution to $2I(x_2^*:x)$ by the additional moment constraints imposed by H_2 but not by H_1, is distributed asymptotically as chi-square with $m_2 - m_1$ degrees of freedom. The foregoing analysis of information can be extended in an obvious manner to a string of nested hypotheses, one successively implying the next.

B. An Iterative Computer Algorithm: Single Sample

For convenience let us first consider the single sample case. We shall then consider an extension of the concepts to the k-sample case. The single sample algorithm is a special case of the k-sample algorithm, but it will be helpful first to consider the single sample case in detail (see Dempster, 1971). We now present an iterative algorithm for the single sample case, essentially a Newton-Raphson type procedure, which provides the MDI estimate with the minimum modified chi-square estimate (see Section VII) as a by-product. For mathematical background see Appendix. Different algorithms are given in Chapter 6.

The following sequence of equalities and statements defines notation and indicates the iterative computations.

(20) $\underset{\sim}{C}\underset{\sim}{p} = \underset{\sim}{\theta}$, $\underset{\sim}{C} = \begin{bmatrix} \underset{\sim}{C}_1 \\ \underset{\sim}{C}_2 \end{bmatrix}$, $\underset{\sim}{C}_1$ is $1 \times \Omega$, $\underset{\sim}{C}_2$ is $r \times \Omega$,

$\underset{\sim}{\theta} = \begin{bmatrix} 1 \\ \underset{\sim}{\theta}^* \end{bmatrix}$, $\underset{\sim}{\theta}^*$ is $r \times 1$,

(21) $\underset{\sim}{C}\underset{\sim}{x} = N\underset{\sim}{\phi}$, $\underset{\sim}{\phi} = \begin{bmatrix} 1 \\ \hat{\underset{\sim}{\theta}} \end{bmatrix}$, $\hat{\underset{\sim}{\theta}}$ is $r \times 1$, $\underset{\sim}{x}$ is an $\Omega \times 1$ matrix of observations, $N = \sum_{\Omega} x(\omega)$,

(22) D_x is an $\Omega \times \Omega$ diagonal matrix of the observations in lexicographic order,

(23) $\underset{\sim}{S} = \underset{\sim}{C}\underset{\sim}{D}_x\underset{\sim}{C}' = \begin{bmatrix} \underset{\sim}{S}_{11} & \underset{\sim}{S}_{12} \\ \underset{\sim}{S}_{21} & \underset{\sim}{S}_{22} \end{bmatrix}$, $\underset{\sim}{S}_{11}$ is 1×1, $\underset{\sim}{S}'_{21} = \underset{\sim}{S}_{12}$ is $1 \times r$, $\underset{\sim}{S}_{22}$ is $r \times r$,

(24) $\underset{\sim}{S}_{22.1} = \underset{\sim}{S}_{22} - \underset{\sim}{S}_{21}\underset{\sim}{S}_{11}^{-1}\underset{\sim}{S}_{12}$,

(25) $\underset{\sim}{\Delta} = N\underset{\sim}{\theta} - N\underset{\sim}{\phi} = \begin{bmatrix} 0 \\ N\underset{\sim}{\theta}^* - N\hat{\underset{\sim}{\theta}} \end{bmatrix} = \begin{bmatrix} 0 \\ \underset{\sim}{d} \end{bmatrix}$,

$\underset{\sim}{d} = N\underset{\sim}{\theta}^* - N\hat{\underset{\sim}{\theta}}$ is $r \times 1$,

(26) $\underset{\sim}{t}^{(j)} = \underset{\sim}{S}_{22.1}^{-1(j)}\underset{\sim}{d}^{(j)}$, $j = 0, 1, 2, \ldots$.

Let $\underset{\sim}{\ell n}\, y$ denote an $\Omega \times 1$ matrix and $\underset{\sim}{\ell n}\, x$ the $\Omega \times 1$ matrix of $\ell n\, x(1), \ldots, \ell n\, x(\Omega)$, where $x(1), \ldots, x(\Omega)$ are the original observations.

(27) $\underset{\sim}{(tau)}^{(j+1)} = \underset{\sim}{(tau)}^{(j)} + t^{(j)}$, $\underset{\sim}{(tau)}^{(j)} \equiv 0$ for $j = 0$, $j = 0, 1, 2, \ldots$,

(28) $\ln y^{(j)} = \ln x + C_2'(\text{tau})^{(j)}$, $j = 1, 2, \ldots$,

(29) $y^{(j)}(1), \ldots, y^{(j)}(\Omega)$, $j = 1, 2, \ldots$,

(30) $L^{(j)} = \ln (N/y^{(j)}(1) + \ldots + y^{(j)}(\Omega))$, $j = 1, 2, \ldots$,

(31)
$$\begin{aligned} \ln x^{(j)}(1) &= L^{(j)} + \ln y^{(j)}(1) \\ &\vdots \\ \ln x^{(j)}(\Omega) &= L^{(j)} + \ln y^{(j)}(\Omega) \end{aligned} \qquad j = 1, 2, \ldots ,$$

(32) $x^{(j)}(1), \ldots, x^{(j)}(\Omega)$, $j = 1, 2, \ldots$.

In step (26), $j = 0$ corresponds to initial values computed in steps (21) to (25) using the original observations, and $j = 1, 2, \ldots$ correspond to iterating the procedures in steps (21) to (25) however using the iterate values

$$x^{(j)}(1), \ldots, x^{(j)}(\Omega) \text{ from step (32).}$$

Note that in step (28) $\ln x$ is always composed of the original observations.

The iteration is continued until the maximum value of the absolute values of the differences between desired constraints and computed constraints is less than a specified small value. The final iterated value $x^{(j)}$ is the MDI estimate x^* and $2I(x^*:x)$ can be computed. It is asymptotically distributed as chi-square with r degrees of freedom. The matrix $S_{22.1}^{-1}$ for the last iterate is the asymptotic covariance matrix of the taus which are the estimated natural parameters of x^*.

If the minimum modified chi-square estimates and the minimum modified chi-square value are desired (see Section VII) one computes in terms of the initial (observed) values

(33) $$\lambda = (CD_xC')^{-1}\Delta = S^{-1}\Delta = \begin{bmatrix} S^{12}d \\ S_{22.1}^{-1}d \end{bmatrix},$$

(34) $\mu = C'\lambda = C'(CD_x C')^{-1}\Delta,$

(35) $\tilde{x} = x + D_x \mu = x + D_x C'(CD_x C')^{-1}\Delta,$

(36) $X^2 = \Delta'\lambda = \Delta'(CD_x C')^{-1}\Delta = d'S_{22.1}^{-1}d.$

The $\tilde{x}$ in (35) are the minimum modified chi-square estimates and X^2 in (36) is the value of the minimum modified chi-square (see Section VII) with r degrees of freedom and is a quadratic approximation to $2I(x^*:x)$. Note that X^2 in (36) can be calculated without first getting $\tilde{x}$.

IV. NUMERICAL EXAMPLE (FISHER'S DATA)

We shall illustrate the single sample algorithm by using Fisher's data as given in Ex. 2 of Section II.

$$x = \begin{bmatrix} 1997 \\ 906 \\ 904 \\ 32 \end{bmatrix}, \quad \ln x = \begin{bmatrix} 7.599401 \\ 6.809039 \\ 6.806829 \\ 3.465736 \end{bmatrix}, \quad D_x = \begin{bmatrix} 1997 & 0 & 0 & 0 \\ 0 & 906 & 0 & 0 \\ 0 & 0 & 904 & 0 \\ 0 & 0 & 0 & 32 \end{bmatrix}$$

$$Cx = N\phi = \begin{bmatrix} 1 & 1 & 1 & 1 \\ 1 & 1 & 0 & 0 \\ 1 & 0 & 1 & 0 \end{bmatrix} \begin{bmatrix} 1997 \\ 906 \\ 904 \\ 32 \end{bmatrix} = \begin{bmatrix} 3839 \\ 2903 \\ 2901 \end{bmatrix}, \quad N = 3839,$$

$$S = CD_x C' = \begin{bmatrix} 3839 & 2903 & 2901 \\ 2903 & 2903 & 1997 \\ 2901 & 1997 & 2901 \end{bmatrix},$$

$$S_{22.1} = \begin{bmatrix} 2903 & 1997 \\ 1997 & 2901 \end{bmatrix} - \begin{bmatrix} 2195.2094 & 2193.6971 \\ 2193.6971 & 2192.1857 \end{bmatrix}$$

$$= \begin{bmatrix} 707.7906 & -196.6971 \\ -196.6971 & 708.8143 \end{bmatrix},$$

$$\Delta = \begin{bmatrix} 0 \\ d \end{bmatrix}, \quad d = N\theta^* - N\hat{\theta} = \begin{bmatrix} -23.75 \\ -21.75 \end{bmatrix},$$

$$\underset{\sim}{S}_{22.1}^{-1} = \begin{bmatrix} .00153 & .00042 \\ .00042 & .00153 \end{bmatrix},$$

$$\underset{\sim}{t}^{(0)} = \begin{bmatrix} .00153 & .00042 \\ .00042 & .00153 \end{bmatrix} \begin{bmatrix} -23.75 \\ -21.75 \end{bmatrix} = \begin{bmatrix} -.045472 \\ -.043253 \end{bmatrix},$$

$$\underset{\sim\sim\sim}{(tau)}^{(1)} = \begin{bmatrix} -.045472 \\ -.043253 \end{bmatrix}, \quad \underset{\sim}{C}_2'\underset{\sim\sim\sim}{(tau)}^{(1)} = \begin{bmatrix} -.088725 \\ -.045472 \\ -.043253 \\ 0 \end{bmatrix},$$

$$\ell n\ y^{(1)}(1) = 7.599401 - 0.088725 = 7.510676,$$
$$\ell n\ y^{(1)}(2) = 6.809039 - 0.045472 = 6.763567,$$
$$\ell n\ y^{(1)}(3) = 6.806829 - 0.043253 = 6.763576,$$
$$\ell n\ y^{(1)}(4) = 3.465736 - 0 = 3.465736,$$

$$y^{(1)}(1) = 1827.448, \quad y^{(1)}(3) = 865.733,$$
$$y^{(1)}(2) = 865.733, \quad y^{(1)}(4) = 32.000,$$

$$y^{(1)}(1) + \dots + y^{(1)}(4) = 3590.914,$$

$$L^{(1)} = \ell n\ (3839/3590.914) = 0.066805,$$

$$\ell n\ x^{(1)}(1) = 0.066805 + 7.510676 = 7.577481, \quad x^{(1)}(1) = 1953.701,$$
$$\ell n\ x^{(1)}(2) = 0.066805 + 6.763567 = 6.830372, \quad x^{(1)}(2) = 925.535,$$
$$\ell n\ x^{(1)}(3) = 0.066805 + 6.763576 = 6.830381, \quad x^{(1)}(3) = 925.543,$$
$$\ell n\ x^{(1)}(4) = 0.066805 + 3.465736 = 3.532541, \quad x^{(1)}(4) = 34.211,$$

$$X^2 = (-23.75, -21.75) \begin{bmatrix} -.045472 \\ -.043253 \end{bmatrix} = 2.021.$$

Retaining two decimal places we take

$$x^*(1) = 1953.71 = x^*(11),$$
$$x^*(2) = 925.54 = x^*(12), \quad x^*(1.) = 2879.25,$$
$$x^*(3) = 925.54 = x^*(21), \quad x^*(.1) = 2879.25,$$
$$x^*(4) = 34.21 = x^*(22).$$

Since $\underset{\sim}{d}^{(1)} = \begin{bmatrix} 0 \\ 0 \end{bmatrix}$, $\underset{\sim}{t}^{(1)} = \underset{\sim}{S}_{22.1}^{-1(1)} \underset{\sim}{d}^{(1)} = \underset{\sim}{0}$ and there will be no change in the estimate by further iteration.

$$2I(x^*:x) = 2(1953.71 \ \ell n \ (1953.71/1997) + 925.54 \ \ell n \ (925.54/906)$$
$$+ 925.54 \ \ell n \ (925.54/904) + 34.21 \ \ell n \ (34.21/906))$$
$$= 2(-42.8174 + 19.7492 + 21.7946 + 2.2846) = 2.022.$$

For Fisher's data, the hypothesis is a consequence of genetic theory and the moment constraints must be consistent therewith. However, to illustrate the procedure, we shall consider the derivation of sets of 95% confidence intervals both for the moment parameters and the natural parameters.

To retain the constraint $x^*(1.) = x^*(.1)$ let us solve the quadratic approximation to $2I(x^*:x)$ (we recall that $2I(x^*:x)$ has 2 D.F.)

$$(y - 2903)^2(0.001531) + (y - 2901)^2(0.001529)$$
$$+ 2(y - 2903)(y - 2901)(0.000425) = 6$$

for y. It is found that $y = 2941.166, 2862.835$. Similarly, one may solve the quadratic equation

$$707.7906 \ \tau_1^2 + 708.8143 \ \tau_2^2 - 2(196.6971)\tau_1 \ \tau_2 = 6$$

subject to the condition $\tau_2 = \tau_1 + 0.00222$ for τ_1. It is found that $\tau_1 = -0.077912, 0.075263$ and $\tau_2 = -0.075692, 0.077483$.

Four MDI estimates were computed, two satisfying the moment constraints determined above and two with values of the natural parameters as determined from the quadratic. In addition a number of other MDI estimates were computed for different values of the moment constraints or natural parameters. We summarize results obtained in the following tabulation

$x^*(1.) = x^*(.1)$	$x^*(1.)/N$	τ_1	τ_2	$2I(x^*:x)$
2862.84	0.746	-0.077335	-0.075126	5.977
2941.17	0.766	+0.075839	+0.078048	6.017
2862.55	0.746	-0.077912	-0.075692	6.070
2940.88	0.766	+0.075263	+0.077483	5.932
2879.25	0.75	-0.045462	-0.043252	2.022
2898.45	0.755	-0.008041	-0.005831	0.050
2840.86	0.74	-0.119890	-0.117680	14.551
2928.24	0.763	+0.050339	+0.052559	2.700
2821.17	0.735	-0.156941	-0.154730	25.091

V. GENERAL FORMULATION AND MDI ANALYSIS (k-SAMPLES)

The extension of the previous single sample discussion to the case of k-samples makes use of an approach due to Gokhale (1973). Consider the k discrete spaces Ω_i, $i = 1, 2, \dots, k$, where we designate the points or cells of Ω_i by $\omega_i(j)$, $j = 1, 2, \dots, \Omega_i$ $\underset{\sim}{\omega}_i' = (\omega_i(1), \dots, \omega_i(\Omega_i))$. We use Ω_i to represent both the space and the number of cells in it. Let $\underset{\sim}{p}_i' = (p_i(\omega_i(1)), \dots, p_i(\omega_i(\Omega_i)))$, $i = 1, \dots, k$, be k sets of probability distributions defined respectively over Ω_i, $i = 1, 2, \dots, k$. Let $\underset{\sim}{p}' = (\underset{\sim}{p}_1', \dots, \underset{\sim}{p}_k')$ be a $1 \times \Omega$ matrix, where $\Omega = \Omega_1 + \Omega_2 + \dots + \Omega_k$. Let P be the collection of all such matrices $\underset{\sim}{p}'$. For a given $\underset{\sim}{\pi}' = (\underset{\sim}{\pi}_1', \dots, \underset{\sim}{\pi}_k') \in P$ and $\underset{\sim}{p}' \in P$ the generalized discrimination information is given by

$$(37)\quad I(\underset{\sim}{p}:\underset{\sim}{\pi}) = \sum_i w_i \sum_{j=1}^{\Omega_i} p_i(\omega_i(j)) \ln(p_i(\omega_i(j))/\pi_i(\omega_i(j))),$$

where the constants w_i are known and are such that $\sum_i w_i = 1$, $0 < w_i < 1$. Let us denote the elements (points or cells) of $\Omega = \Omega_1 + \Omega_2 + \dots + \Omega_k$ by $\omega(ij)$, $i = 1, \dots, k$, $j = 1, \dots, \Omega_i$, so that $\omega(i1), \dots, \omega(i\Omega_i)$ are the components of Ω belonging to Ω_i.

The MDI estimate is the value of $\underset{\sim}{p}$ which minimizes the generalized discrimination information in (37) over the family of $\underset{\sim}{p}$'s which satisfy the linearly independent moment constraints

$$\underset{\sim}{B}\,\underset{\sim}{p} = \underset{\sim}{\theta}, \tag{38}$$

where $\underset{\sim}{B}$ is $(k + r) \times \Omega$, $\underset{\sim}{p}$ is $\Omega \times 1$, $\underset{\sim}{\theta}$ is $(k + r) \times 1$ and the rank of $\underset{\sim}{B}$ is $(k + r) \leq \Omega$. We shall now transform the problem to a canonical form similar to that of the single sample case. Note that for ECP the matrix $\underset{\sim}{\pi}$ consists of the k observed distributions.

A. Canonical Form

Let $\underset{\sim}{W}_i$ be an $\Omega_i \times \Omega_i$ diagonal matrix with diagonal elements w_i, and

$$\underset{\sim}{W} = \begin{bmatrix} \underset{\sim}{W}_1 & 0 & \cdots & 0 \\ 0 & \underset{\sim}{W}_2 & \cdots & 0 \\ \cdot & \cdot & \cdots & \cdot \\ 0 & 0 & \cdots & \underset{\sim}{W}_k \end{bmatrix}, \ \Omega \times \Omega, \tag{39}$$

$$\underset{\sim}{P} = \underset{\sim}{W}\underset{\sim}{p},\ \underset{\sim}{P}' = (P(\omega(11)), \ldots, P(\omega(1\Omega_1)), \ldots, P(\omega(k1)), \ldots, P(\omega(k,\Omega_k))) \tag{40}$$

$$\underset{\sim}{\Pi} = \underset{\sim}{W}\underset{\sim}{\pi},\ \underset{\sim}{\Pi}' = (\Pi(\omega(11)), \ldots, \Pi(\omega(k\Omega_k))), \tag{41}$$

$$\underset{\sim}{C} = \underset{\sim}{B}\underset{\sim}{W}^{-1},\ \underset{\sim}{C} \text{ is } (k + r) \times \Omega,\ \underset{\sim}{W}^{-1} = \underset{\sim}{V}. \tag{42}$$

We note that

$$\sum_{\Omega} P(\omega) = \sum_{j=1}^{\Omega_1} w_1 p_1(\omega_1(j)) + \ldots + \sum_{j=1}^{\Omega_k} w_k p_k(\omega_k(j)) = w_1 + \ldots + w_k = 1, \tag{43}$$

$$\sum_{\Omega} \Pi(\omega) = \sum_{j=1}^{\Omega_1} w_1 \pi_1(\omega_1(j)) + \ldots + \sum_{j=1}^{\Omega_k} w_k \pi_k(\omega_k(j)) = w_1 + \ldots + w_k = 1, \tag{44}$$

(45) $$I(\underset{\sim}{p}:\underset{\sim}{\pi}) = \sum_{i=1}^{k} \sum_{j=1}^{\Omega_i} w_i p_i(\omega_i(j)) \ell n\, (w_i p_i(\omega_i(j))/w_i \pi_i(\omega_i(j)))$$

$$= \sum_{\Omega} P(\omega) \ell n(P(\omega)/\Pi(\omega)) = I(P:\Pi),$$

(46) $$\underset{\sim}{B}\underset{\sim}{p} = \underset{\sim}{B}\underset{\sim}{W}^{-1}\underset{\sim}{W}\underset{\sim}{p} = \underset{\sim}{C}\underset{\sim}{P} = \underset{\sim}{\theta}.$$

In terms of the canonical transformation the k-sample problem may now be formulated as finding the MDI estimate $P^*(\omega)$ minimizing

(47) $$I(P:\Pi) = \sum_{\Omega} P(\omega) \ell n\, (P(\omega)/\Pi(\omega)),$$

subject to the linearly independent constraints

(48) $$\underset{\sim}{C}\underset{\sim}{P} = \underset{\sim}{\theta},$$

where $\underset{\sim}{C}$ is $(k + r) \times \Omega$, $\underset{\sim}{P}$ is $\Omega \times 1$, $\underset{\sim}{\theta}$ is $(k + r) \times 1$ and the rank of $\underset{\sim}{C}$ is $(k + r) \leqq \Omega$. Paralleling the discussion of the single sample case, with appropriate modifications, we denote the elements of the matrix $\underset{\sim}{C}$ by $c_i(\omega)$, $i = 1, \ldots, k, k + 1, \ldots, k + r$, $\omega = 11, \ldots, 1\Omega_1, \ldots, k1, \ldots, k\Omega_k$. We may write (48) as

(49) $$\sum_{\Omega} c_i(\omega)P(\omega) = \theta_i,\ i = 1, \ldots, k, k + 1, \ldots, k + r.$$

We shall usually take the elements of the B-matrix $b_i(\omega_i(j)) = 1$, $j = 1, \ldots, \Omega_i$, $i = 1, \ldots, k$, and zero otherwise, that is, the natural constraint for each sample, so that

(50) $$c_i(\omega) = v_i \text{ for } \omega = i1, \ldots, i\Omega_i,\ v_i = 1/w_i,$$

$$= 0 \text{ otherwise},\ i = 1, 2, \ldots, k,$$

$$\theta_i = 1,\ i = 1, 2, \ldots, k.$$

The exponential family or multiplicative model for the MDI estimate is $P^*(\omega) = \exp(L_1c_1(\omega) + \ldots + L_kc_k(\omega) + \tau_1c_{k+1}(\omega) + \ldots + \tau_rc_{k+r}(\omega))\Pi(\omega)$ and the loglinear representation is

(51) $$\ell n(P^*(\omega)/\Pi(\omega)) = L_1 c_1(\omega) + \ldots + L_k c_k(\omega) + \tau_1 c_{k+1}(\omega) + \ldots + \tau_r c_{k+r}(\omega),\ \omega = 11, \ldots, k\ \Omega_k.$$

The L's are normalizing constants and the τ's are the natural parameters of interest.

We now partition the matrices as follows:

(52) $$\underset{\sim}{C} = \begin{bmatrix} \underset{\sim}{C}_1 \\ \underset{\sim}{C}_2 \end{bmatrix},\ \text{where } \underset{\sim}{C}_1 \text{ is } k \times \Omega,\ \underset{\sim}{C}_2 \text{ is } r \times \Omega,$$

(53) $$\underset{\sim}{\theta} = \begin{bmatrix} \underset{\sim}{1} \\ \underset{\sim}{\theta}^* \end{bmatrix},\ \text{where } \underset{\sim}{1} \text{ is a } k \times 1 \text{ matrix of 1's, } \underset{\sim}{\theta}^* \text{ is } r \times 1,$$

$$\text{that is, } \underset{\sim}{C}_1\underset{\sim}{P} = \underset{\sim}{1},\ \underset{\sim}{C}_2\underset{\sim}{P} = \underset{\sim}{\theta}^*.$$

If there are k samples corresponding to $\Omega_1, \ldots, \Omega_k$, where the sum of the observations in the i-th sample is N_i and $N = N_1 + N_2 + \ldots + N_k$, then we set $w_i = N_i/N$,

(54) $$x^*(\omega) = NP^*(\omega),$$

(55) $$x(\omega) = N\Pi(\omega),$$

(56) $$\sum_{\Omega} x(\omega) = \sum_{j=1}^{\Omega_1} N_1\, x_1(j)/N_1 + \ldots + \sum_{j=1}^{\Omega_k} N_k\, x_k(j)/N_k$$

$$= N_1 + \ldots + N_k = N.$$

The MDI statistic is

(57) $$2I(x^*:x) = 2NI(P^*:\Pi) = 2 \sum_{\Omega} x^*(\omega)\ell n(x^*(\omega)/x(\omega)),$$

which is asymptotically distributed as chi-square with r degrees of freedom if the observed values are consistent with the null hypothesis (38) or (48), or the estimate $x^*(\omega)$. If we set

(58) $$\underset{\sim}{C}N\underset{\sim}{\Pi} = \underset{\sim}{C}\underset{\sim}{x} = N\underset{\sim}{\phi},\ \underset{\sim}{\phi} = \begin{bmatrix} \underset{\sim}{1} \\ \underset{\sim}{\hat{\theta}} \end{bmatrix},\ \text{where } \underset{\sim}{x} \text{ is } \Omega \times 1,$$

$$\underset{\sim}{1} \text{ is a } k \times 1 \text{ matrix of 1's, } \underset{\sim}{\theta} \text{ is } r \times 1,$$

then a quadratic approximation to $2I(x^*:x)$ is given by the minimum modified chi-square with r D.F.,

(59) $$\chi^2 = (N\underset{\sim}{\theta}^* - N\hat{\underset{\sim}{\theta}})'\underset{\sim}{S}_{22.1}^{-1}(N\underset{\sim}{\theta}^* - N\hat{\underset{\sim}{\theta}}),$$

where

(60) $$\underset{\sim}{S} = \underset{\sim}{C}\underset{\sim}{D}_x\underset{\sim}{C}' = \begin{bmatrix} \underset{\sim}{C}_1\underset{\sim}{D}_x\underset{\sim}{C}_1' & \underset{\sim}{C}_1\underset{\sim}{D}_x\underset{\sim}{C}_2' \\ \underset{\sim}{C}_2\underset{\sim}{D}_x\underset{\sim}{C}_1' & \underset{\sim}{C}_2\underset{\sim}{D}_x\underset{\sim}{C}_2' \end{bmatrix} = \begin{bmatrix} \underset{\sim}{S}_{11} & \underset{\sim}{S}_{12} \\ \underset{\sim}{S}_{21} & \underset{\sim}{S}_{22} \end{bmatrix},$$

and $\underset{\sim}{S}_{11}$ is k x k, $\underset{\sim}{S}_{21}' = \underset{\sim}{S}_{12}$ is k x r, $\underset{\sim}{S}_{22}$ is r x r,

$\underset{\sim}{S}_{22.1} = \underset{\sim}{S}_{22} - \underset{\sim}{S}_{21}\underset{\sim}{S}_{11}^{-1}\underset{\sim}{S}_{12}$.

B. An Iterative Computer Algorithm: k-Samples

For convenience (computer-wise) we shall use n_i for Ω_i and n for Ω, that is, $n = n_1 + n_2 + \ldots + n_k$, where n_i is the number of cells in the i-th sample whose total number of observations is N_i. The following sequence of equalities and statement defines notation and indicates the iterative computations

(61) $\underset{\sim}{C}\underset{\sim}{P} = \underset{\sim}{\theta}$, $\underset{\sim}{C} = \begin{bmatrix} \underset{\sim}{C}_1 \\ \underset{\sim}{C}_2 \end{bmatrix}$, $\underset{\sim}{C}_1$ is k x n, $\underset{\sim}{C}_2$ is r x n,

$\underset{\sim}{\theta} = \begin{bmatrix} \underset{\sim}{1} \\ \underset{\sim}{\theta}^* \end{bmatrix}$, $\underset{\sim}{1}$ is a k x 1 matrix of ones, $\underset{\sim}{\theta}^*$ is r x 1,

(62) $\underset{\sim}{C}\underset{\sim}{x} = N\underset{\sim}{\phi}$, $\underset{\sim}{\phi} = \begin{bmatrix} \underset{\sim}{1} \\ \hat{\underset{\sim}{\theta}} \end{bmatrix}$, $\underset{\sim}{1}$ is a k x 1 matrix of ones, $\hat{\underset{\sim}{\theta}}$ is r x 1,

(63) $\underset{\sim}{D}_x$ is an n x n diagonal matrix of observations in lexicographic order,

(64) $\underset{\sim}{S} = \underset{\sim}{C}\underset{\sim}{D}_x\underset{\sim}{C}' = \begin{bmatrix} \underset{\sim}{S}_{11} & \underset{\sim}{S}_{12} \\ \underset{\sim}{S}_{21} & \underset{\sim}{S}_{22} \end{bmatrix}$, $\underset{\sim}{S}_{11}$ is k x k, $\underset{\sim}{S}_{21}' = \underset{\sim}{S}_{12}$ is k x r,

$\underset{\sim}{S}_{22}$ is r x r,

(65) $$\underset{\sim}{S}_{22.1} = \underset{\sim}{S}_{22} - \underset{\sim}{S}_{21}\underset{\sim}{S}_{11}^{-1}\underset{\sim}{S}_{12},$$

(66) $\underset{\sim}{\Delta} = \underset{\sim}{N\theta} - \underset{\sim}{N\phi} = \begin{bmatrix} \underset{\sim}{0} \\ \underset{\sim}{d} \end{bmatrix}$, $\underset{\sim}{0}$ is a k x 1 matrix of zeros,

$\underset{\sim}{d} = \underset{\sim}{N\theta^*} - \underset{\sim}{N\hat{\theta}}$ is r x 1,

(67) $$\underset{\sim}{t}^{(j)} = \underset{\sim}{S}_{22.1}^{-1(j)}\underset{\sim}{d}^{(j)}, \ j = 0, 1, 2, \ldots .$$

Let $\underset{\sim\sim}{\ell n}$ y denote an n x 1 matrix and $\underset{\sim\sim}{\ell n}$ x the n x 1 matrix of $\ell n\ x(1), \ldots, \ell n\ x(n)$,

(68) $\underset{\sim\sim\sim}{(tau)}^{(j+1)} = \underset{\sim\sim\sim}{(tau)}^{(j)} + t^{(j)}$, $\underset{\sim\sim\sim}{(tau)}^{(j)} \equiv 0$ for $j = 0$,

$j = 0, 1, 2, \ldots ,$

(69) $$\underset{\sim\sim}{\ell n}\ y^{(j)} = \underset{\sim\sim}{\ell n}\ x + \underset{\sim}{C}_2'\underset{\sim\sim\sim}{(tau)}^{(j)}, \ j = 1, 2, \ldots ,$$

(70) $$y^{(j)}(1), \ldots , y^{(j)}(n), \ j = 1, 2, \ldots ,$$

(71)

$S_1^{(j)} = \sum_{\Omega_1} y^{(j)}(\omega)$ for ω the n_1 values in the first set,

$\vdots \quad \vdots \quad \vdots$

$S_k^{(j)} = \sum_{\Omega_k} y^{(j)}(\omega)$ for ω the n_k values in the $k^{\underline{th}}$ set,

(72) $$v_h L_h^{(j)} = M_h^{(j)} = \ell n\ N_h/S_h^{(j)}, \ h = 1, 2, \ldots , k,$$

(73) $\ell n\ x^{(j)}(\omega) = M_h^{(j)} + \ell n\ y^{(j)}(\omega)$, for ω in set
$h = 1, 2, \ldots , k, \ j = 1, 2, \ldots ,$

(74) $$x^{(j)}(1), \ldots , x^{(j)}(n), \ j = 1, 2, \ldots .$$

In step (67), j = 0 corresponds to initial values computed in steps (62) to (66) using $\underset{\sim}{x}$ and j = 1, 2, ... correspond to iterating the procedures in (62) to (66) however using the iterate values $x^{(j)}(1), \ldots , x^{(j)}(n)$ in step (74). Note that in step (69) $\underset{\sim\sim}{\ell n}$ x is <u>always</u> composed of the initial values $\underset{\sim}{x}$. The iteration is continued until the maximum of the absolute values of the differences between desired constraints and computed constraints is less than a specified small value. The final iterated value $\underset{\sim}{x}^{(j)}$ is the MDI estimate $\underset{\sim}{x}^*$ and $2I(x^*:x)$ can be computed. It is asymptotically distributed as chi-square with r degrees of freedom. The matrix $\underset{\sim}{S}_{22.1}^{-1}$ for the last iterate is the asymptotic covariance matrix of the taus which are the estimated natural parameters of $\underset{\sim}{x}^*$.

If the minimum modified chi-square estimates are desired (see Section VII) one computes in terms of the initial (observed) values,

$$(75) \quad \underset{\sim}{\lambda} = (\underset{\sim}{C}\underset{\sim}{D}_x\underset{\sim}{C}')^{-1}\underset{\sim}{\Delta} = \underset{\sim}{S}^{-1}\underset{\sim}{\Delta} = \begin{bmatrix} \underset{\sim}{S}^{12}\underset{\sim}{d} \\ \underset{\sim}{S}_{22.1}^{-1}\underset{\sim}{d} \end{bmatrix},$$

$$(76) \quad \underset{\sim}{u} = \underset{\sim}{C}'\underset{\sim}{\lambda} = \underset{\sim}{C}'(\underset{\sim}{C}\underset{\sim}{D}_x\underset{\sim}{C}')^{-1}\underset{\sim}{\Delta},$$

$$(77) \quad \tilde{\underset{\sim}{x}} = \underset{\sim}{x} + \underset{\sim}{D}_x\underset{\sim}{\mu} = \underset{\sim}{x} + \underset{\sim}{D}_x\underset{\sim}{C}'(\underset{\sim}{C}\underset{\sim}{D}_x\underset{\sim}{C}')^{-1}\underset{\sim}{\Delta},$$

$$(78) \quad X^2 = \underset{\sim}{\Delta}'\underset{\sim}{\lambda} = \underset{\sim}{\Delta}'(\underset{\sim}{C}\underset{\sim}{D}_x\underset{\sim}{C}')^{-1}\underset{\sim}{\Delta} = \underset{\sim}{d}'\underset{\sim}{S}_{22.1}^{-1}\underset{\sim}{d}.$$

The $\tilde{\underset{\sim}{x}}$ in (77) are the minimum modified chi-square estimates and X^2 in (78) is the value of the minimum modified chi-square (see Section VII) with r degrees of freedom and is a quadratic approximation to $2I(x^*:x)$. Note that X^2 in (78) can be computed without getting $\tilde{\underset{\sim}{x}}$.

VI. AN ELEMENTARY EXAMPLE

We present an elementary example illustrating the $2I(x^*:x)$ quadratic approximation using the several sample approach. Suppose we have observed two binomial samples

$$x(11),\ x(12),\ x(11) + x(12) = N_1,$$

$$x(21),\ x(22),\ x(21) + x(22) = N_2,$$

and we want to test the null hypothesis that $p(11) = p(21)$. The set up corresponding to $\underset{\sim}{B}\underset{\sim}{p} = \underset{\sim}{\theta}$ is

	11	12	21	22	
ω:	1	2	3	4	θ
	1	1	0	0	1
	0	0	1	1	1
	1	0	-1	0	0

Using $v_1 = 1/w_1 = N/N_1$, $v_2 = 1/w_2 = N/N_2$, $N = N_1 + N_2$, the transformation to $\underset{\sim}{C}\underset{\sim}{P} = \underset{\sim}{\theta}$ is

	11	12	21	22	
ω:	1	2	3	4	θ
	v_1	v_1	0	0	1
	0	0	v_2	v_2	1
	v_1	0	$-v_2$	0	0

With $\underset{\sim}{D}_x$ a diagonal matrix with entries x(ij) in lexicographic order, we compute

$$\underset{\sim}{C}\underset{\sim}{D}_x\underset{\sim}{C}' = \begin{bmatrix} v_1^2(x(11) + x(12)) & 0 & v_1^2 x(11) \\ 0 & v_2^2(x(21) + x(22)) & -v_2^2 x(21) \\ v_1^2 x(11) & -v_2^2 x(21) & v_1^2 x(11) + v_2^2 x(21) \end{bmatrix}$$

We now find

$$S_{22.1} = v_1^2 x(11) + v_2^2 x(21) - v_1^2 x^2(11)/N_1 - v_2^2\, x^2(21)/N_2$$

$$= v_1^2 x(11)(1 - x(11)/N_1) + v_2^2 x(21)(1 - x(21)/N_2).$$

But

$$\underset{\sim}{d} = 0 - (v_1 x(11) - v_2 x(21)),$$

hence $X^2 = \underset{\sim}{d}'\underset{\sim}{S}_{22.1}^{-1}\,\underset{\sim}{d}$ is

$$X^2 = \frac{(v_1x(11) - v_2x(21))^2}{v_1^2x(11)(1 - x(11)/N_1) + v_2^2x(21)(1 - x(21)/N_2)}$$

$$= \frac{(Nx(11)/N_1 - Nx(21)/N_2)^2}{(N^2/N_1^2)x(11)(1 - x(11)/N_1) + (N^2/N_2^2)x(21)(1 - x(21)/N_2)}$$

$$= \frac{(\hat{p}(11) - \hat{p}(21))^2}{\hat{p}(11)\hat{q}(11)/N_1 + \hat{p}(21)\hat{q}(21)/N_2}$$

$$\hat{p}(11) = x(11)/N_1, \ \hat{q}(11) = 1 - \hat{p}(11),$$

$$\hat{p}(21) = x(21)/N_2, \ \hat{q}(21) = 1 - \hat{p}(21).$$

See Kullback (1959, 311), Snedecor and Cochran (1967, 496).

VII. MINIMUM MODIFIED CHI-SQUARE ESTIMATION

We shall use the same notation as in Section III. For minimum modified chi-square estimation we want the value of $p(\omega)$ which minimizes the modified chi-square

$$\chi_1^2 = \sum_{\Omega} (p(\omega) - \pi(\omega))^2/\pi(\omega) \tag{79}$$

subject to the constraints $\underset{\sim}{C}\underset{\sim}{p} = \underset{\sim}{\theta}$ (see (14)).

Differentiating (79) with respect to $p(\omega)$ and using Lagrange multipliers we have

$$(\tilde{p}(\omega) - \pi(\omega))/\pi(\omega) - \lambda_0c_0(\omega) - \dots - \lambda_rc_r(\omega) = 0, \tag{80}$$

$$\omega = 1, \dots, \Omega.$$

If we set $\xi(\omega) = (\tilde{p}(\omega) - \pi(\omega))/\pi(\omega)$, $\underset{\sim}{\xi}' = (\xi(1), \dots, \xi(\Omega))$, $\underset{\sim}{\lambda}' = (\lambda_0, \dots, \lambda_r)$, then (80) may be written as (matrix notation)

$$\underset{\sim}{\xi} = \underset{\sim}{C}'\underset{\sim}{\lambda}, \tag{81}$$

or

$$\tilde{\underset{\sim}{p}} = \underset{\sim}{\pi} + \underset{\sim}{D}_{\pi}\underset{\sim}{C}'\underset{\sim}{\lambda}, \tag{82}$$

where $\tilde{p}' = (\tilde{p}(1), \ldots, \tilde{p}(\Omega))$, $\pi' = (\pi(1), \ldots, \pi(\Omega))$, and D_π is the $\Omega \times \Omega$ diagonal matrix with main diagonal $\pi(1), \ldots, \pi(\Omega)$. If we set (see (21))

$$(83) \qquad C\pi = \phi, \ \phi' = (1, \hat{\theta}'),$$

then from (82) we get

$$(84) \qquad C(\tilde{p} - \pi) = \theta - \phi = CD_\pi C'\lambda,$$

or

$$(85) \qquad \lambda = (CD_\pi C')^{-1}(\theta - \phi),$$

that is

$$(86) \qquad \tilde{p} = \pi + D_\pi C'(CD_\pi C')^{-1}(\theta - \phi),$$

or, with $\tilde{x} = N\tilde{p}$, $x = N\pi$,

$$(87) \qquad \tilde{x} = x + D_x C'(CD_x C')^{-1}(N\theta - N\phi),$$

where $D_x = ND_\pi$. Since

$$(88) \quad \min \chi_1^2 = \sum_\Omega (\tilde{x}(\omega) - x(\omega))^2/x(\omega)$$

$$= (D_x^{-1/2}(\tilde{x} - x))'(D_x^{-1/2}(\tilde{x} - x))$$

and from (87)

$$(89) \quad D_x^{-1/2}(\tilde{x} - x) = D_x^{1/2}C'(CD_x C')^{-1}(N\theta - N\phi),$$

we have

$$(90) \quad \min \chi_1^2 = (N\theta - N\phi)'(CD_x C')^{-1}(N\theta - N\phi).$$

Using the notation of (23), (36)

$$(91) \qquad (CD_x C')^{-1} = S^{-1} = \begin{bmatrix} S^{11} & S^{12} \\ S^{21} & S_{22.1}^{-1} \end{bmatrix},$$

$$(92) \qquad (N\underset{\sim}{\theta} - N\underset{\sim}{\phi})' = (0, N\underset{\sim}{\theta}^* - N\underset{\sim}{\hat{\theta}})',$$

hence

$$(93) \quad \min \chi_1^2 = (N\underset{\sim}{\theta}^* - N\underset{\sim}{\hat{\theta}})'\underset{\sim}{S}_{22.1}^{-1}(N\underset{\sim}{\theta}^* - N\underset{\sim}{\hat{\theta}}),$$

that is, the right-hand side of (19).

Chapter 6

COMPUTER ALGORITHMS

I. INTRODUCTION

Two programs for the analysis of up to four-way contingency tables using iterative marginal fitting were prepared by Dr. H. H. Ku and Mrs. Ruth Varner of the Statistical Engineering Laboratory, National Bureau of Standards, in 1967 (Ku and Kullback (1968)).

In the summer of 1970 the first version of The George Washington University program CONTAB using iterative marginal fitting and providing broader capabilities was written by Dr. C. T. Ireland filling a need carefully defined by Professor Solomon Kullback. This program has gone through several revisions and expansions, including some practical innovations suggested by Dr. Frederick Scheuren of the Long Range Research Branch of the Social Security Administration. The current version NEWTAB has been prepared by John Nolan.

A modification of CONTAB was prepared in 1972 by Dr. Marian Fisher, now at National Heart and Lung Institute, Bethesda, Maryland, 20014. The current version of this program is in use at The George Washington University Computer Center as CONTABMOD. It provides as output, in addition to the MDI estimates and their logarithms, the design matrices, values of the natural parameters (taus), and their covariance matrix.

KULLITR, GOKHALE, and DARRAT are computer programs applicable to ECP, as well as to ICP. These programs are of a general nature, not restricted to marginal fitting techniques and were compiled by Dr. John C. Keegel beginning in September 1973 and are in use at The George Washington University Computer Center, The University of California, Riverside and other university, government and research installations. KULLITR uses a Newton-Raphson type technique implementing the iterative algorithm described in Chapter 5. GOKHALE uses a method of steepest descent. The initial approximation to the MDI estimates must satisfy the constraints. DARRAT uses the Darrat-Ratcliff generalized iterative scaling technique. DARRAT uses the least storage of the three algorithms and does not use matrix inversion; however convergence is the slowest. KULLITR requires the most storage but converges fastest. It is recommended that KULLITR be used whenever possible. These programs are written using PL/1. A magnetic tape reel of the programs and manual may be obtained by writing to Secretary, Department of Statistics, The George Washington University, Washington, D.C. 20052. There will be a minimal charge to cover the tapes, manual, handling and postage.

For ready reference we include a chi-square table. We first illustrate the iterative marginal fitting or scaling algorithm for the two-way marginals of a three-way table.

II. CHI-SQUARE TABLE

Probability $\chi^2 \geq \chi$

Significance Level

Degrees of Freedom	0.5	0.1	.05	.01	.005	.001	.0001	.00001	.000001	
1	0.5	2.7	3.8	6.6	7.9	10.8	15.1	19.5	23.9	1
2	1.4	4.6	6.0	9.2	10.6	13.8	18.4	23.0	27.6	2
3	2.4	6.2	7.8	11.3	12.8	16.3	21.1	25.9	30.7	3
4	3.4	7.8	9.5	13.3	14.9	18.5	23.5	28.5	33.4	4
5	4.4	9.2	11.1	15.1	16.8	20.5	25.7	30.9	35.9	5
6	5.3	10.6	12.6	16.8	18.5	22.5	27.9	33.1	38.3	6
7	6.3	12.0	14.1	18.5	20.3	24.3	29.9	35.3	40.5	7
8	7.3	13.4	15.5	20.1	22.0	26.1	31.8	37.3	42.7	8
9	8.3	14.7	16.9	21.7	23.6	27.9	33.7	39.3	44.8	9
10	9.3	16.0	18.3	23.2	25.2	29.6	35.6	41.3	46.9	10
11	10.3	17.3	19.7	24.7	26.8	31.3	37.4	43.2	48.9	11
12	11.3	18.6	21.0	26.2	28.3	32.9	39.1	45.1	50.8	12
13	12.3	19.8	22.4	27.7	29.8	34.5	40.9	46.9	52.8	13
14	13.3	21.1	23.7	29.1	31.3	36.1	42.6	48.7	54.6	14
15	14.3	22.3	25.0	30.6	32.8	37.7	44.3	50.5	56.5	15
16	15.3	23.5	26.3	32.0	34.3	39.3	45.9	52.2	58.3	16
17	16.3	24.8	27.6	33.4	35.7	40.8	47.6	54.0	60.1	17
18	17.3	26.0	28.9	34.8	37.2	42.3	49.2	55.7	61.9	18
19	18.3	27.2	30.1	36.2	38.6	43.8	50.8	57.4	63.7	19
20	19.3	28.4	31.4	37.6	40.0	45.3	52.4	59.0	65.4	20
21	20.3	29.6	32.7	38.9	41.4	46.8	54.0	60.7	67.1	21
22	21.3	30.8	33.9	40.3	42.8	48.3	55.5	62.3	68.9	22
23	22.3	32.0	35.2	41.6	44.2	49.7	57.1	64.0	70.6	23
24	23.3	33.2	36.4	43.0	45.6	51.2	58.6	65.6	72.2	24
25	24.3	34.4	37.7	44.3	46.9	52.6	60.1	67.2	73.9	25
26	25.3	35.6	38.9	45.6	48.3	54.1	61.7	68.8	75.6	26
27	26.3	36.7	40.1	47.0	49.6	55.5	63.2	70.4	77.2	27
28	27.3	37.9	41.3	48.3	51.0	56.9	64.7	71.9	78.8	28
29	28.3	39.1	42.6	49.6	52.3	58.3	66.2	73.5	80.4	29
30	29.3	40.3	43.8	50.9	53.7	59.7	67.6	75.0	82.0	30
31	30.3	41.4	45.0	52.2	55.0	61.1	69.1	76.6	83.6	31
32	31.3	42.6	46.2	53.5	56.3	62.5	70.6	78.1	85.2	32
33	32.3	43.8	47.4	54.8	57.7	63.9	72.0	79.6	86.8	33
34	33.3	44.9	48.6	56.1	59.0	65.3	73.5	81.1	88.4	34
35	34.3	46.1	49.8	57.3	60.3	66.6	74.9	82.6	90.0	35
48	47.3	60.9	65.2	73.7	77.0	84.0	93.2	101.7	109.7	48
49	48.3	62.0	66.3	74.9	78.2	85.4	94.6	103.1	111.1	49
63	62.3	77.8	82.5	92.0	95.7	103.4	113.5	122.7	131.4	63
80	79.3	96.6	101.9	112.3	116.3	124.8	135.3	145.8	155.1	80
99	98.3	117.4	123.2	134.6	139.0	143.2	160.1	170.8	180.8	99

III. ITERATION, MARGINAL FITTING ALGORITHM

The values of the MDI estimate p* can be computed by an iterative scheme which adjusts the initial values to satisfy successively the given marginal constraints. For a three-way r x c x d table when all two-way marginals p(ij.), p(i.k), p(.jk) are given, the iteration cycles through

$$\text{(1)}\quad \begin{aligned} p^{(3n+1)}(ijk) &= \frac{p(ij.)}{p^{(3n)}(ij.)}\, p^{(3n)}(ijk) \\ p^{(3n+2)}(ijk) &= \frac{p(i.k)}{p^{(3n+1)}(i.k)}\, p^{(3n+1)}(ijk) \\ p^{(3n+3)}(ijk) &= \frac{p(.jk)}{p^{(3n+2)}(.jk)}\, p^{(3n+2)}(ijk) \quad, n = 0, 1, \ldots \end{aligned}$$

where $p^{(0)}(ijk)$ may be 1/rcd or $p_1^*(ijk) = p(i..)p(.j.)p(..k)$. For a four-way r x s x t x u table when all three-way marginals p(ijk.), p(ij.ℓ), p(i.kℓ), p(.jkℓ) are given the iteration cyles through

$$\begin{aligned} p^{(4n+1)}(ijk\ell) &= \frac{p(ijk.)}{p^{(4n)}(ijk.)}\, p^{(4n)}(ijk\ell) \\ p^{(4n+2)}(ijk\ell) &= \frac{p(ij.\ell)}{p^{(4n+1)}(ij.\ell)}\, p^{(4n+1)}(ijk\ell) \\ p^{(4n+3)}(ijk\ell) &= \frac{p(i.k\ell)}{p^{(4n+2)}(i.k\ell)}\, p^{(4n+2)}(ijk\ell) \\ p^{(4n+4)}(ijk\ell) &= \frac{p(.jk\ell)}{p^{(4n+3)}(.jk\ell)}\, p^{(4n+3)}(ijk\ell) \end{aligned}$$

where $p^{(0)}(ijk\ell)$ may be 1/rstu or the MDI estimates $p_1^*(ijk\ell)$, $p_2^*(ijk\ell)$, respectively fitting one-way or two-way marginals.

It can be shown that the iteration converges to a unique p*.

Although the above iterations have been given in terms of probabilities in practice it has been found more convenient not to divide everything by n and the iterations are carried out using observed or estimated occurrences, $n\pi(ijk\ell) = n/rstu$, $x(i\ldots)$, $x(ij..)$, etc., $x^*(ijk\ell) = np^*(ijk\ell)$. In certain cases when the MDI estimates can be given explicitly as products of marginals the iteration is completed after the first cycle, for example, the MDI estimate fitting the observed one-way marginals is $x_1^*(ijk\ell) = x(i\ldots)x(.j..)x(..k.)x(\ldots\ell)/n^3$. Usually 5 to 7 cycles have been found to be sufficient to obtain agreement between desired and computed marginals to within 0.001 when more than one cycle is required.

It may be helpful to elaborate somewhat the iterative algorithm given in (1) in terms of occurrences as follows:

1. Start with $\overset{(0)}{x}(ijk) = n/rcd$.

2. Compute the marginals $\overset{(0)}{x}(ij.)$.

3. Adjust $\overset{(0)}{x}(ijk)$ by the ratios of the observed marginals $x(ij.)$ to computed marginals $\overset{(0)}{x}(ij.)$. The adjusted entries are $\overset{(1)}{x}(ijk)$.

4. Compute the marginals $\overset{(1)}{x}(i.k)$.

5. Adjust $\overset{(1)}{x}(ijk)$ by the ratios of the observed marginals $x(i.k)$ to the computed marginals $\overset{(1)}{x}(i.k)$. The adjusted entries are $\overset{(2)}{x}(ijk)$.

6. Compute the marginals $\overset{(2)}{x}(.jk)$.

7. Adjust $\overset{(2)}{x}(ijk)$ by the ratios of the observed marginals

$x(.jk)$ to the computed marginals $x^{(2)}(.jk)$. The adjusted entries are $x^{(3)}(ijk)$ and one cycle is completed.

8. Continue the procedure from steps (2) through (7) above using $x^{(3)}(ijk)$ as the starting entries.

9. Continue the process until the three sets of computed marginals agree to within the specified tolerance.

We shall illustrate the iterative algorithm (1) with Cochran's data (1954) for the 2 x 2 x 3 Table 1.

TABLE 1

Data on number of mothers with previous infant losses

Birth Order		Number of mothers with losses	Number of mothers with no losses
2	Problem	x(111) = 20	x(121) = 82
	Control	x(211) = 10	x(221) = 54
3 - 4	Problem	x(112) = 26	x(122) = 41
	Control	x(212) = 16	x(222) = 30
5+	Problem	x(113) = 27	x(123) = 22
	Control	x(213) = 14	x(223) = 23

The sets of observed marginals are

x(ij.)		x(.jk)			x(i.k)		
73	145	30	42	41	102	67	49
40	107	136	71	45	64	46	37

We shall find the values of the MDI estimate $x_2^*(ijk)$ fitting these marginals.

Using $x^{(0)}(ijk) = 365/12 = 30.416$ the values on pp. 222-3 Table 2 are obtained. After the first cycle, the resemblance between $x^{(3)}(ijk)$ and the final values $x_2^*(ijk)$ is already evident, and the tolerance requirement of 0.001 is met after 5 cycles.

IV. GOKHALE

GOKHALE is a computer program that implements an algorithm presented by Gokhale (1972). The algorithm may be characterized as a method of steepest descent, that calculates the MDI estimate minimizing

$$I = \Sigma p_t \ell n \ (p_t/\pi_t) \tag{1}$$

subject to the linearly independent constraints

$$\underset{\sim}{C}\underset{\sim}{p} = \underset{\sim}{\theta}. \tag{2}$$

This is achieved by examining only estimates that satisfy the constraints (2) and following the gradient of (1) in the direction of steepest descent. The procedure converges to the MDI estimate.

We now describe the algorithm which the GOKHALE program implements. The notation and formulation of Chapter 5, Section V is used. Consider the problem of minimizing

$$I(P:\Pi) = \Sigma_t \ P_t \ \ell n(P_t/\Pi_t) \tag{1}$$

with respect to the linearly independent constraints

$$\underset{\sim}{C}\underset{\sim}{P} = \underset{\sim}{\theta}. \tag{2}$$

Note that $\underset{\sim}{C} = \underset{\sim}{B}\underset{\sim}{W}^{-1}$, $\underset{\sim}{P} = \underset{\sim}{W}\underset{\sim}{p}$ and $\underset{\sim}{\Pi} = \underset{\sim}{W}\underset{\sim}{\pi}$. There exists a unique $\underset{\sim}{P}^*$ which minimizes (1) and satisfies

$$\underset{\sim}{\ell n} \ P^* = \underset{\sim}{\ell n} \ \Pi + \underset{\sim}{C}'\underset{\sim}{\lambda} \tag{3}$$

where $\underset{\sim}{\ell n}$ a denotes $(\ell n \ a_1, \ \ldots \ , \ \ell n \ a_t)'$, and $\underset{\sim}{\lambda}$ is a vector of Lagrangian multipliers. Let

$$\underset{\sim}{C}^+ = \underset{\sim}{C}'(\underset{\sim}{C}\underset{\sim}{C}')^{-1} \text{ and } \underset{\sim}{R} = \underset{\sim}{C}^+\underset{\sim}{C}. \tag{4}$$

Then equation (3) is equivalent to

$$(\underset{\sim}{I} - \underset{\sim}{R})(\underset{\sim}{\ell n} \ P^* - \underset{\sim}{\ell n} \ \Pi) = \underset{\sim}{0}. \tag{5}$$

The symmetric and idempotent matrix $\underset{\sim}{R}$ projects vectors of dimension equal to that of $\underset{\sim}{P}$ onto the space spanned by rows of $\underset{\sim}{C}$. Let

$$U = \{z: \; C^{+}\theta + (I - R)\, z > 0\},$$

where for a vector x, $x > 0$ denotes that every element of x is positive. Then for every $z \in U$, $C^{+}\theta + (I - R)z$ is a solution of (2). Conversely, for every probability vector P which satisfies (2), there exists a $z \in U$ such that $P = C^{+}\theta + (I - R)z$. The first assertion is easy to verify and the second follows by setting $z = P$ and noting the $C^{+}\theta = RP$. Consider (1) as a function of z defined over U. Then $I(z) \geqq 0$, the gradient $G(z)$ of $I(z)$ at z is

$$G(z) = (I - R)(\ell n\, P(z) - \ell n\, \Pi) \tag{6}$$

and the Hessian of I at z is

$$H(z) = (I - R)(\Delta(P(z)))^{-1}(I - R), \tag{7}$$

where $\Delta(b)$ denotes a diagonal matrix with elements of the vector b in the diagonal. $I - R$ is positive definite since it is idempotent, so that I is a convex function of z over the convex set U. Thus for a z_o satisfying $G(z_o) = 0$, $I(z)$ assumes its minimum over U. In fact, $G(z_o) = 0$ implies that the corresponding $P(z_o)$ satisfies (5) in view of (6).

At the s-th iteration the algorithm uses a vector $z(s)$ in U and the corresponding $P(s) = C^{+}\theta + (I - R)z(s)$. If

$$|G(s)| = |G[z(s)]| \leqq \varepsilon, \tag{8}$$

where $\varepsilon > 0$ is chosen according to the required accuracy, the procedure is terminated and P^* is set equal to $P(s)$. If (8) does not hold, the direction $D(s)$ of maximum rate of decrease in $I(z)$ at $z(s)$ is obtained by norming $(-G(s))$. A positive constant $c(s)$, sufficiently small, is then found such that with $z(s + 1) = z(s) + c(s)D(s)$ and $P(s + 1) = C^{+}\theta(I - R)z(s + 1)$

$$P(s + 1) > 0 \tag{9}$$

and

$$I(s + 1) < I(s). \tag{10}$$

The $(s + 1)$-th iteration is started with $\underset{\sim}{z}(s + 1)$ and $\underset{\sim}{P}(s + 1)$. One way of finding $c(s)$ is to first set it equal to unity. It is repeatedly doubled until one of (9) or (10) is violated. If (9) or (10) do not hold with $c(s) = 1$, it is repeatedly halved until they do.

Consider now the choice of $\underset{\sim}{z}(1)$ and $\underset{\sim}{P}(1)$. If some $\hat{\underset{\sim}{P}} > 0$ is known to satisfy (2), we set $\underset{\sim}{P}(1) = \underset{\sim}{z}(1) = \hat{\underset{\sim}{P}}$. If not, $\hat{\underset{\sim}{P}}$ can be found easily, though by trial and error, by several methods. One method is to compute $\underset{\sim}{q}(\underset{\sim}{\xi}) = \Delta(\underset{\sim}{\xi})\underset{\sim}{C}'[\underset{\sim}{C}\Delta(\underset{\sim}{\xi})\underset{\sim}{C}']^{-1}\underset{\sim}{\theta}$ for a positive probability vector $\underset{\sim}{\xi}$ and set $\hat{\underset{\sim}{P}} = \underset{\sim}{q}(\underset{\sim}{\xi})$ if the latter is positive. Another method is to check whether $\underset{\sim}{C}^{+}\underset{\sim}{\theta} + (\underset{\sim}{I} - \underset{\sim}{R})\underset{\sim}{\xi}$ is positive and, if so, set it equal to $\underset{\sim}{P}(1)$. Usually, putting $\underset{\sim}{\xi}$ equal to the observed probability vector gives the desired value of $\hat{\underset{\sim}{P}}$. In fact, then $\underset{\sim}{q}(\underset{\sim}{\xi})$ is the minimum modified chi-square estimate of $\underset{\sim}{P}$ subject to (2), which minimizes $\Sigma(P_t - \xi_t)^2\xi_t$, while $\underset{\sim}{C}^{+}\underset{\sim}{\theta} + (\underset{\sim}{I} - \underset{\sim}{R})\underset{\sim}{\xi}$ minimizes the Euclidean distance between $\underset{\sim}{P}$ and $\underset{\sim}{\xi}$. As such, these $\hat{\underset{\sim}{P}}$ serve as good starting points for the iterations.

V. DARRAT

The generalized iterative scaling procedure described by Darroch and Ratcliff (1972), extends the Deming-Stephan iterative scaling algorithm to cases in which the design matrix does not consist only of zeros and ones. A discussion of the procedure and the proof of the convergence of the iteration are to be found in the cited reference. Examples of its use are given in Chapter 8 Example 7 and Kullback and Cornfield (1976).

For convenience as a frame of reference we present the generalized iterative scaling algorithm as given by Darroch and Ratcliff. Let I be a finite set and let $\underset{\sim}{p} = [p(i); i \in I, p(i) \geq 0, \Sigma\, p(i) = 1]$ be a probability function on I. Suppose that p is a member of a family of distributions satisfying the constraints

$$(1) \qquad \Sigma\, b_{si}p(i) = k_s, \; s = 1, 2 \ldots d, \; \Sigma\, p(i) = 1$$

where for all s there exist $i \in I$ such that $b_{si} \neq 0$. The constraints in (1) may be reformulated into the equivalent canonical form

$$\Sigma\, a_{ri}p(i) = h_r,\ r = 1, 2, \ldots c,$$
(2)
$$a_{ri} \geqq 0,\quad \sum_{r=1}^{c} a_{ri} = 1,\ h_r > 0,\quad \sum_{r=1}^{c} h_r = 1,$$

by defining

$$a_{si} = t_s(u_s + b_{si}),\ \text{all } i,$$
(3)
$$h_s = t_s(u_s + k_s),\ s = 1, 2, \ldots, d,$$

where $u_s \geqq 0$, $t_s > 0$ are chosen to make

$$a_{si} \geqq 0 \text{ and } \sum_{s=1}^{d} a_{si} \leqq 1 \text{ for all } i \in I.$$

If $\sum_{s=1}^{d} a_{si} = 1$ for all i define c = d, otherwise define $c = d + 1$ and let $a_{ci} = 1 - \sum_{s=1}^{d} a_{si}$, $h_c = 1 - \sum_{s=1}^{d} h_s$.

Now let $\tilde{\pi} = [\pi(i),\ i \in I,\ \pi(i) > 0,\ \Sigma\, \pi(i) \leqq 1]$ be a subprobability function on I. The MDI estimate p*(i), i ε I, is that member of the family $\tilde{p}$ satisfying the constraints (2) and minimizing

(4) $$I(\tilde{p}:\tilde{\pi}) = \Sigma\, p(i)\, \ell n\, (p(i)/\pi(i))$$

and is given by

(5) $$\ell n\, (p^*(i)/\pi(i)) = \sum_{r=1}^{c} a_{ri}\tau_r,$$

where the τ_r are parameters to be determined so that p*(i) satisfies the constraints (2). The values of p*(i) may be determined by the convergent iteration

$$(6) \quad p^{(n+1)}(i) = p^{(n)}(i) \prod_{r=1}^{c} \left(\frac{h_r}{h_r^{(n)}}\right)^{a_{ri}}, \quad n = 0, 1, 2, \ldots$$

$$p^{(0)}(i) = \pi(i), \quad h_r^{(n)} = \Sigma\, a_{ri} p^{(n)}(i).$$

We remark that if we use the notation $i = \omega$, $I = \Omega$, $k = \theta$, $b_{si} = b_s(\omega)$, then the constraints in (1) above are the same as the constraints (38) in Chapter 5, Section V.

TABLE 2

ijk	(1) x(ijk)	(2) x(ijk)	(3) x(ijk)	(4) x(ijk)	(5) x(ijk)
111	24.333	34.156	19.869	20.079	20.427
211	13.333	17.415	10.130	9.941	9.679
121	48.333	67.844	80.633	80.184	81.573
221	35.667	46.585	55.367	55.791	54.321
112	24.333	22.435	26.959	27.244	27.019
212	13.333	12.517	15.041	14.759	14.937
122	48.333	44.564	40.540	40.314	39.981
222	35.667	33.483	30.369	30.693	31.063
113	24.333	16.408	25.410	25.677	25.074
213	13.333	10.067	15.590	15.299	15.805
123	48.333	32.592	24.639	24.502	23.926
223	35.667	26.932	20.361	20.516	21.195

(6) x(ijk)		(14) x(ijk)	Fitted Data (15) x(ijk)	Original Data x(ijk)
20.355	...	20.503	20.503	20
9.645	...	9.497	9.497	10
81.637	...	81.497	81.497	82
54.363	...	54.503	54.503	54
27.047	...	27.213	27.213	26
14.953	...	14.787	14.787	16
39.957	...	39.787	39.787	41
31.043	...	31.213	31.213	30
25.148	...	25.284	25.284	27
15.852	...	15.716	15.716	14
23.862	...	23.716	23.716	22
21.138	...	21.284	21.284	23

Chapter 7

NO INTERACTION ON A LINEAR SCALE

I. NO LINEAR INTERACTION

In this chapter and the one following, we consider more examples of ECP. This chapter is devoted to an important special case: the hypotheses of no interaction on a linear scale in contingency tables of the one response many factors type. In such tables the data can be looked upon as a collection of as many multinomial experiments as there are factor-level combinations and each experiment has a number of cells equal to the levels of the response variable. We also relate quadratic approximations of the MDI statistics to Wald-type statistics and Neyman's modified chi-square. Multiplicative and additive definitions of no interaction in contingency tables are discussed in Darroch (1974).

In this chapter we also indicate a way of reducing categories in a contingency table with the inherent qualities of the observed data least affected. In the next chapter we shall also see illustrations of treating a contingency table as a collection of multinomial experiments by the application of the k-sample procedure to ICP.

One formulation of a no linear interaction hypothesis is that the cell probabilities can be expressed as <u>linear</u> functions of parameters which are structurally less complex. For example, in a r x s x t x u table in which the first variable is a response variable and the rest are factors, an hypothesis of no third order linear interaction may be expressed as

$$(1)\quad H_0:\ p(ijk\ell) = \mu(ijk.) + \mu(ij.\ell) + \mu(i.k\ell)$$

where the $p(ijk\ell)$ are subject to the (natural) constraints

$$(2)\quad \Sigma_{i=1}^{r} p(ijk\ell) = 1, \text{ for each combination } (jk\ell),\ j = 1, \ldots, s,$$

$$k = 1, \ldots, t \text{ and } \ell = 1, \ldots, u.$$

The functions μ (which can be regarded as parameters) depend only on the indicated indices. The hypothesis H_0 is equivalent to the following $(r - 1)(s - 1)(t - 1)(u - 1)$ constraints in addition to those in (2)

$$(3)\quad \begin{aligned} &p(ijk\ell) - p(isk\ell) - p(ijt\ell) + p(ist\ell) \\ &- p(ijku) + p(isku) + p(ijtu) - p(istu) = 0, \\ &i = 1, \ldots, (r - 1),\ j = 1, \ldots, (s - 1), \\ &k = 1, \ldots, (t - 1),\ \ell = 1, \ldots, (u - 1). \end{aligned}$$

It may be seen from (1) that for fixed i each μ-function on the righthand side depends on two variables only, thus showing absence of third order interaction. A similar hypothesis of no third and second order interaction on a linear scale can be formulated by letting

$$(4)\quad p(ijk\ell) = \mu(ij..) + \mu(i.k.) + \mu(i..\ell).$$

In addition to (2), the above hypothesis implies the constraints (3) and a set of $(r - 1)(s - 1)(t - 1) + (r - 1)(s - 1)(u - 1) + (r - 1)(t - 1)(u - 1) + (s - 1)(t - 1)(u - 1)$ constraints, obtained by fixing each index in turn. For example, fixing the index ℓ we note that

$$(5)\quad p(ijk\ell) - p(isk\ell) - p(ijt\ell) + p(ist\ell) = 0,$$

$$i = 1, \dots, (r-1),\ j = 1, \dots, (s-1),$$
$$k = 1, \dots, (t-1),$$

so that we have $(r - 1)(s - 1)(t - 1)$ constraints corresponding to fixed ℓ.

The foregoing discussion shows that the no linear interaction hypotheses can be formulated as linear constraints on the underlying probabilities, written in matrix notation as $\underset{\sim}{B}\underset{\sim}{p} = \underset{\sim}{\theta}$. It is possible to obtain MDI estimates of cell frequencies and test various hypotheses and sub-hypotheses. If the hypotheses are nested the MDI statistic for the stronger hypothesis (which imposes more constraints) can be analyzed into two components, one measuring the deviation between the observed distribution and the MDI estimate under the weaker hypothesis and the other measuring the deviation between the MDI estimates under the two hypotheses. This feature of the MDI statistics is not enjoyed by the chi-square type or Wald-type statistics used by many authors.

For the sake of clarity of presentation, we shall restrict examples to the hypotheses of no linear <u>second order</u> interaction in a 2 x 2 x 2 table and in a 4 x 2 x 2 table. This enables us to compare results with Bhapkar and Koch (1968), who have viewed two sets of data as of the one response many factors type. The analysis of a 2 x 2 x 2 table shows how the use of an approximation in the MDI statistic leads to a statistic used by Bhapkar and Koch (1968). The 4 x 2 x 2 table is analysed under two hypotheses of no linear interaction of second order, illustrating the analysis of information mentioned in the preceding paragraph.

II. THREE-WAY TABLE: GENERAL FORMULATION

For a three-way r x s x t table in which the first variable is a response and the other two variables are factors, one formulation of no linear second-order interaction is given by

$$\text{(6)} \qquad H_o\colon\ p(ijk) = \mu(i..) + \mu(ij.) + \mu(i.k),$$

$$i = 1, \dots, r,\ j = 1, \dots, s,\ k = 1, \dots, t,$$

where the p(ijk) are subject to the constraints

(7) $\Sigma_{i=1}^{r} p(ijk) = 1$, for each fixed pair (jk),

and the parameters μ depend only on the indicated indices. The hypothesis H_0 is equivalent to the following $(r - 1)(s - 1)(t - 1)$ constraints in addition to those in (7)

(8) $p(ijk) - p(ijt) - p(isk) + p(ist) = 0$,

$i = 1, \ldots, (r - 1), j = 1, \ldots, (s - 1)$,

$k = 1, \ldots, (t - 1)$.

Writing

$$\underset{\sim}{p} = (p(111), p(211), \ldots, p(r11), p(112), \ldots, p(rst))',$$

where the (jk) indices are in lexicographic order, the constraints (7) and (8) can be expressed as

(9) $$\underset{\sim}{B}\underset{\sim}{p} = \underset{\sim}{\theta}$$

where the vector $\underset{\sim}{\theta}$ consists of the first st elements equal to unity and the remaining elements equal to zero. This is illustrated in the examples below.

III. THE 2 X 2 X 2 TABLE

Consider the probabilities of the 2 x 2 x 2 contingency Table 1.

TABLE 1

		B j = 1		β j = 2	
		C k = 1	γ k = 2	C k = 1	γ k = 2
i = 1	A	p(111)	p(112)	p(121)	p(122)
i = 2	α	p(211)	p(212)	p(221)	p(222)

The experimental procedure selects a fixed number of observations under the four possible combinations of the factors (B,β), (C,γ) and determines the number of occurrences of (A,α) for each case. In effect then the procedure is examining four binomials with

(10) $p(1jk) + p(2jk) = 1, \ j = 1, 2, \ k = 1, 2.$

The corresponding observed values are shown in Table 2. It is desired to test whether the observed values are consistent with a null hypothesis of no interaction on a linear scale, that is

$$H_0: \ p(111) - p(112) = p(121) - p(122)$$
(11)
$$\text{or } p(111) - p(112) - p(121) + p(122) = 0.$$

TABLE 2

	j = 1		j = 2	
	k = 1	k = 2	k = 1	k = 2
i = 1	x(111)	x(112)	x(121)	x(122)
i = 2	x(211)	x(212)	x(221)	x(222)
	x(.11)	x(.12)	x(.21)	x(.22)

We shall determine MDI estimates for the cell entries subject to the null hypothesis and compare the estimated and observed values. The estimated table is given in Table 3 where the λ's are to be determined.

TABLE 3

	j = 1		j = 2	
	k = 1	k = 2	k = 1	k = 2
i = 1	$x(111) + \lambda_1$	$x(112) + \lambda_2$	$x(121) + \lambda_3$	$x(122) + \lambda_4$
i = 2	$x(211) - \lambda_1$	$x(212) - \lambda_2$	$x(221) - \lambda_3$	$x(222) - \lambda_4$
	x(.11)	x(.12)	x(.21)	x(.22)

We shall use the principle of MDI estimation and thus determine the λ's which minimize the discrimination information

$$
\begin{aligned}
&(x(111) + \lambda_1)\ln \frac{x(111) + \lambda_1}{x(111)} + (x(211) - \lambda_1)\ln \frac{x(211) - \lambda_1}{x(211)} \\
&+ (x(112) + \lambda_2)\ln \frac{x(112) + \lambda_2}{x(112)} + (x(212) - \lambda_2)\ln \frac{x(212) - \lambda_2}{x(212)} \\
(12) \quad &+ (x(121) + \lambda_3)\ln \frac{x(121) + \lambda_3}{x(121)} + (x(221) - \lambda_3)\ln \frac{x(221) - \lambda_3}{x(221)} \\
&+ (x(122) + \lambda_4)\ln \frac{x(122) + \lambda_4}{x(122)} + (x(222) - \lambda_4)\ln \frac{x(222) - \lambda_4}{x(222)} \\
&+ \tau \left(\frac{x(111) + \lambda_1}{x(.11)} - \frac{x(112) + \lambda_2}{x(.12)} - \frac{x(121) + \lambda_3}{x(.21)} + \frac{x(122) + \lambda_4}{x(122)} \right),
\end{aligned}
$$

where τ is a Lagrange undetermined multiplier and the constraint (11) is reflected by the condition

$$
(13) \quad \frac{x(111) + \lambda_1}{x(.11)} - \frac{x(112) + \lambda_2}{x(.12)} - \frac{x(121) + \lambda_3}{x(.21)} + \frac{x(122) + \lambda_4}{x(.22)} = 0.
$$

Differentiating (12) with respect to $\lambda_1, \ldots, \lambda_4$ leads to the equations

$$
(14) \quad
\begin{aligned}
&\ln \frac{x(111) + \lambda_1}{x(111)} - \ln \frac{x(211) - \lambda_1}{x(211)} + \frac{\tau}{x(.11)} = 0, \\
&\ln \frac{x(112) + \lambda_2}{x(112)} - \ln \frac{x(212) - \lambda_2}{x(212)} - \frac{\tau}{x(.12)} = 0, \\
&\ln \frac{x(121) + \lambda_3}{x(121)} - \ln \frac{x(221) - \lambda_3}{x(221)} - \frac{\tau}{x(.21)} = 0, \\
&\ln \frac{x(122) + \lambda_4}{x(122)} - \ln \frac{x(222) - \lambda_4}{x(222)} + \frac{\tau}{x(.22)} = 0.
\end{aligned}
$$

There are a number of different iterative approaches to determine the solution to (14) but our interest here is to examine the relation of an approximate solution to other proposed methods.

Assuming that the ratios of the λ's to the observed values are small, we use the approximations

$$\ell n \frac{x(111) + \lambda_1}{x(111)} \approx \frac{\lambda_1}{x(111)}, \quad \ell n \frac{x(211) - \lambda_1}{x(211)} \approx - \frac{\lambda_1}{x(211)}, \text{ etc.,}$$

in (14) and get

$$(15) \quad \begin{aligned} &\frac{\lambda_1}{x(111)} + \frac{\lambda_1}{x(211)} + \frac{\tau}{x(.11)} = 0 = \lambda_1 \frac{x(.11)}{x(111)x(211)} + \frac{\tau}{x(.11)}, \\ &\frac{\lambda_2}{x(112)} + \frac{\lambda_2}{x(212)} - \frac{\tau}{x(.12)} = 0 = \lambda_2 \frac{x(.12)}{x(112)x(212)} - \frac{\tau}{x(.12)}, \\ &\frac{\lambda_3}{x(121)} + \frac{\lambda_3}{x(221)} - \frac{\tau}{x(.21)} = 0 = \lambda_3 \frac{x(.21)}{x(121)x(221)} - \frac{\tau}{x(.21)}, \\ &\frac{\lambda_4}{x(122)} + \frac{\lambda_4}{x(222)} + \frac{\tau}{x(.22)} = 0 = \lambda_4 \frac{x(.22)}{x(122)x(222)} + \frac{\tau}{x(.22)}. \end{aligned}$$

From (15) and (13) we have, introducing the notation $x(1jk) = x(.jk)\hat{p}(jk)$, $x(2jk) = x(.jk)\hat{q}(jk)$, $\hat{p}(jk) + \hat{q}(jk) = 1$,

$$(16) \quad \begin{aligned} \lambda_1 &= - \frac{x(111)x(211)}{(x(.11))^2} \tau = - \hat{p}(11)\hat{q}(11)\tau, \\ \lambda_2 &= \frac{x(112)x(212)}{(x(.12))^2} \tau = \hat{p}(12)\hat{q}(12)\tau, \\ \lambda_3 &= \frac{x(121)x(221)}{(x(.21))^2} \tau = \hat{p}(21)\hat{q}(21)\tau, \\ \lambda_4 &= - \frac{x(122)x(222)}{(x(.22))^2} \tau = - \hat{p}(22)\hat{q}(22)\tau, \\ \tau &= \frac{\hat{p}(11) - \hat{p}(12) - \hat{p}(21) + \hat{p}(22)}{\frac{\hat{p}(11)\hat{q}(11)}{x(.11)} + \frac{\hat{p}(12)\hat{q}(12)}{x(.12)} + \frac{\hat{p}(21)\hat{q}(21)}{x(.21)} + \frac{\hat{p}(22)\hat{q}(22)}{x(.22)}}. \end{aligned}$$

Let us write

$$(17) \quad \begin{aligned} &x^*(111) = x(111) + \lambda_1, \; x^*(211) = x(211) - \lambda_1, \\ &x^*(112) = x(112) + \lambda_2, \; x^*(212) = x(212) - \lambda_2, \text{ etc.,} \end{aligned}$$

where the λ's satisfy (14). If we also use the quadratic approximations

$$2\left((x(111)+\lambda_1)\ln\frac{x(111)+\lambda_1}{x(111)}+(x(211)-\lambda_1)\ln\frac{x(211)-\lambda_1}{x(211)}\right) \tag{18}$$

$$\approx \lambda_1^2\left(\frac{1}{x(111)}+\frac{1}{x(211)}\right)=\lambda_1^2\frac{x(.11)}{x(111)x(211)}=\frac{\lambda_1^2}{x(.11)\hat{p}(11)\hat{q}(11)}$$

then we get for the MDI statistic

$$2I(x^*:x) = 2\Sigma\Sigma\Sigma\, x^*(ijk)\ln\,(x^*(ijk)/x(ijk))$$

$$\approx \tau^2\left(\frac{\hat{p}(11)\hat{q}(11)}{x(.11)}+\frac{\hat{p}(12)\hat{q}(12)}{x(.12)}+\frac{\hat{p}(21)\hat{q}(21)}{x(.21)}+\frac{\hat{p}(22)\hat{q}(22)}{x(.22)}\right)$$

$$=\frac{(\hat{p}(11)-\hat{p}(12)-\hat{p}(21)+\hat{p}(22))^2}{\frac{\hat{p}(11)\hat{q}(11)}{x(.11)}+\frac{\hat{p}(12)\hat{q}(12)}{x(.12)}+\frac{\hat{p}(21)\hat{q}(21)}{x(.21)}+\frac{\hat{p}(22)\hat{q}(22)}{x(.22)}} \tag{19}$$

$$=\lambda_1^2\left(\frac{1}{x(111)}+\frac{1}{x(211)}\right)+\lambda_2^2\left(\frac{1}{x(112)}+\frac{1}{x(212)}\right)$$

$$+\ldots+\lambda_4^2\left(\frac{1}{x(122)}+\frac{1}{x(222)}\right).$$

Note that the last value in (19) is the minimum modified Neyman chi-square

$$\chi_1^2 = \Sigma\ (\text{obs}-\text{exp})^2/\text{obs} \tag{20}$$

and indeed the equations in (15) are those to determine the minimum modified chi-square estimates. The next to last value in (19) is the statistic given by Bhapkar and Koch (1968, 116) based on a criterion due to Wald. The square root of this value is the statistic used by Snedecor and Cochran (1967, 496).

In accordance with the principle of MDI estimation the log-linear representation for $x^*(ijk)$ is given graphically as in Figure 1 where the interpretation is

$$\ell n\ (x^*(111)/x(111)) = L_1 - \tau/x(.11),$$

$$\ell n\ (x^*(211)/x(211)) = L_1,$$

(21)
$$\ell n\ (x^*(112)/x(112)) = L_2 + \tau/x(.12),$$

$$\ell n\ (x^*(212)/x(212)) = L_2,$$

$$\cdots\cdots\cdots$$

$$\ell n\ (x^*(222)/x(222)) = L_4.$$

Recalling (17) we see that (21) in fact leads to (14).
If we write

(22)
$$\begin{aligned} \theta^* &= \frac{x^*(111)}{x(.11)} - \frac{x^*(112)}{x(.12)} - \frac{x^*(121)}{x(.21)} + \frac{x^*(122)}{x(.22)} \\ &= p^*(11) - p^*(12) - p^*(21) + p^*(22), \\ \hat{\theta} &= \frac{x(111)}{x(.11)} - \frac{x(112)}{x(.12)} - \frac{x(121)}{x(.21)} + \frac{x(122)}{x(.22)} \\ &= \hat{p}(11) - \hat{p}(12) - \hat{p}(21) + \hat{p}(22), \end{aligned}$$

then as shown in Kullback (1959, 101-106)

(23)
$$2I(x^*:x) \simeq (\theta^* - \hat{\theta})^2/\hat{\sigma}^2,$$

where $\hat{\sigma}^2$ is determined as follows. Let $\underset{\sim}{C}'$ denote the 8 x 5 matrix in Figure 1, that is,

(24)
$$\underset{\sim}{C}' = \begin{bmatrix} 1 & 0 & 0 & 0 & -1/x(.11) \\ 1 & 0 & 0 & 0 & 0 \\ 0 & 1 & 0 & 0 & +1/x(.12) \\ 0 & 1 & 0 & 0 & 0 \\ 0 & 0 & 1 & 0 & +1/x(.21) \\ 0 & 0 & 1 & 0 & 0 \\ 0 & 0 & 0 & 1 & -1/x(.22) \\ 0 & 0 & 0 & 1 & 0 \end{bmatrix}$$

and $\underset{\sim}{D}_x$ the 8 x 8 diagonal matrix with entries x(ijk), that is,

$$
(25) \quad \underset{\sim}{D}_x = \begin{bmatrix}
x(111) & 0 & . & . & . & . & . & 0 \\
0 & x(211) & & & & & & . \\
. & & x(112) & & & & & . \\
. & & & x(212) & & & & . \\
. & & & & x(121) & & & . \\
. & & & & & x(221) & & . \\
. & & & & & & x(122) & . \\
0 & . & . & . & . & . & . & x(222)
\end{bmatrix}
$$

Compute the 5 x 5 matrix $\underset{\sim}{S} = \underset{\sim}{C}\underset{\sim}{D}_x\underset{\sim}{C}'$ and partition it as follows

$$
(26) \qquad \underset{\sim}{S} = \begin{bmatrix} \underset{\sim}{S}_{11} & \underset{\sim}{S}_{12} \\ \underset{\sim}{S}_{21} & \underset{\sim}{S}_{22} \end{bmatrix},
$$

$\underset{\sim}{S}_{11}$ is 4 x 4, $\underset{\sim}{S}_{22}$ is 1 x 1, $\underset{\sim}{S}_{21} = \underset{\sim}{S}'_{12}$ is 1 x 4,

then $\hat{\sigma}^2$ in (23) is given by

$$
(27) \qquad \hat{\sigma}^2 = \underset{\sim}{S}_{22} - \underset{\sim}{S}_{21}\underset{\sim}{S}_{11}^{-1}\underset{\sim}{S}_{12}.
$$

It may be verified that this results in

$$
(28) \qquad \begin{aligned}
\hat{\sigma}^2 &= \frac{x(111)x(211)}{(x(.11))^3} + \frac{x(112)x(212)}{(x(.12))^3} + \frac{x(121)x(221)}{(x(.21))^3} + \frac{x(122)x(222)}{(x(.22))^3} \\
&= \frac{\hat{p}(11)\hat{q}(11)}{x(.11)} + \frac{\hat{p}(12)\hat{q}(12)}{x(.12)} + \frac{\hat{p}(21)\hat{q}(21)}{x(.21)} + \frac{\hat{p}(22)\hat{q}(22)}{x(.22)} .
\end{aligned}
$$

But θ^* in (22) is zero and we see that (23) is indeed the next-to-last value in (19). It is interesting to note that $2I(x^*:x)$ can be approximated without necessarily computing the values of $x^*(ijk)$.

Note now that in order to express the hypothesis H_o of (11) in the form $\underset{\sim}{B}\underset{\sim}{p} = \underset{\sim}{\theta}$, we can let

$$\underset{\sim}{p} = (p(111),\ p(211),\ p(112),\ p(212),\ p(121),\ p(221),\ p(122),\ p(222))',$$

(29)
$$\underset{\sim}{B} = \begin{bmatrix} 1 & 1 & 0 & 0 & 0 & 0 & 0 & 0 \\ 0 & 0 & 1 & 1 & 0 & 0 & 0 & 0 \\ 0 & 0 & 0 & 0 & 1 & 1 & 0 & 0 \\ 0 & 0 & 0 & 0 & 0 & 0 & 1 & 1 \\ 1 & 0 & -1 & 0 & -1 & 0 & 1 & 0 \end{bmatrix}$$

and

(30)
$$\underset{\sim}{\theta} = (1,\ 1,\ 1,\ 1,\ 0)'.$$

FIGURE 1

ijk	L_1	L_2	L_3	L_4	τ
111	1				-1/x(.11)
211	1				
112		1			+1/x(.12)
212		1			
121			1		+1/x(.21)
221			1		
122				1	-1/x(.22)
222				1	

We shall illustrate the preceding discussion with Bartlett's data on root cuttings used also as an example by Snedecor and Cochran (1967), Bhapkar and Koch (1968), Berkson (1972). The following Table 4 from Bartlett (1935), who refers to data from Hoblyn and Palmer, is the result of an experiment designed to investigate the propagation of plum root stocks from root cuttings. There were 240 cuttings for each of the four treatments.

TABLE 4

	At Once $j = 1$		In Spring $j = 2$	
	Long $k = 1$	Short $k = 2$	Long $k = 1$	Short $k = 2$
Dead $i = 1$	84	133	156	209
Alive $i = 2$	156	107	84	31
	240	240	240	240

By using the $\underset{\sim}{B}$ and $\underset{\sim}{\theta}$ defined in (29) and (30), the MDI estimates $x^*(ijk)$ of the cell-frequencies are obtained as

82.885	134.213	157.115	208.443
157.115	105.787	82.885	31.557 .

They agree within round-off errors with those obtained by Berkson (1972). The MDI statistic $2I(x^*:x)$ equals 0.0810 with one D.F.

The value of the natural parameter τ associated with the no linear second-order interaction constraint is $\tau = -0.00512$ and its variance is 0.00032. We note that $(-0.00512)^2/0.00032 = 0.0819$ is a quadratic approximation to $2I(x^*:x)$.

The minimum modified chi-square estimates are

82.883	134.213	157.117	208.448
157.117	105.787	82.883	31.552

and the minimum modified chi-square value is 0.0818.

IV. THE 4 X 2 X 2 TABLE

Analysis of hypotheses of no linear interaction in a 4 x 2 x 2 table is illustrated by Schotz's data Table 5 on drivers in injury producing accidents, taken from Table III of Bhapkar and Koch (1968), who regard accident severity as response and the other two classifications as factors.

Let us ignore for the moment the numerical severity "ridit" scores r_i, $i = 1, \ldots, 4$ and consider the hypothesis of no linear second-order interaction formulated in (8). The B-matrix is

Cell:

111 211 311 411 112 212 312 412 121 221 321 421 122 222 322 422

(31)

$$\underset{\sim}{B} = \begin{matrix} 1 & 1 & 1 & 1 & 0 & 0 & 0 & 0 & 0 & 0 & 0 & 0 & 0 & 0 & 0 & 0 \\ 0 & 0 & 0 & 0 & 1 & 1 & 1 & 1 & 0 & 0 & 0 & 0 & 0 & 0 & 0 & 0 \\ 0 & 0 & 0 & 0 & 0 & 0 & 0 & 0 & 1 & 1 & 1 & 1 & 0 & 0 & 0 & 0 \\ 0 & 0 & 0 & 0 & 0 & 0 & 0 & 0 & 0 & 0 & 0 & 0 & 1 & 1 & 1 & 1 \\ 1 & 0 & 0 & 0 & -1 & 0 & 0 & 0 & -1 & 0 & 0 & 0 & 1 & 0 & 0 & 0 \\ 0 & 1 & 0 & 0 & 0 & -1 & 0 & 0 & 0 & -1 & 0 & 0 & 0 & 1 & 0 & 0 \\ 0 & 0 & 1 & 0 & 0 & 0 & -1 & 0 & 0 & 0 & -1 & 0 & 0 & 0 & 1 & 0 \end{matrix}$$

and

(32) $$\underset{\sim}{\theta} = (1,1,1,1,0,0,0)'.$$

The MDI estimates of cell frequencies are

$x^*(111) = 27.32$	$x^*(211) = 531.50$
$x^*(121) = 932.32$	$x^*(221) = 5535.91$
$x^*(112) = 14.59$	$x^*(212) = 583.70$
$x^*(122) = 734.45$	$x^*(222) = 4884.30$
$x^*(311) = 1359.16$	$x^*(411) = 670.02$
$x^*(321) = 2768.49$	$x^*(421) = 1242.28$
$x^*(312) = 1733.23$	$x^*(412) = 842.48$
$x^*(322) = 2522.48$	$x^*(422) = 1106.77$

The MDI statistic $2I(x^*:x)$ with 3 d.f. is 19.703, which is significant at the 5% level, showing that the data do <u>not</u> support the hypothesis of no linear second-order interaction as given by (8).

It is interesting to examine here the hypothesis of no linear second-order interaction with respect to average "ridits" considered by Bhapkar and Koch (1968). The hypothesis is

$$H_1: \; A_k = \Sigma_{i=1}^{4} r_i[p(i1k) - p(i2k)] = A, \; k = 1,2,$$

where A is a constant. This is equivalent to $A_1 - A_2 = 0$. The

TABLE 5

Driver Group (k)	Accident Severity (i) / "ridit" (r) / Accident Type (j)	Minor
	"ridit" (r)	.05
Lone Driver	Rollover	21
	Non-rollover	996
	Sub-total	1017
Injured Driver with Passengers	Rollover	18
	Non-rollover	679
	Sub-total	697
Total		1714

Moderate	Moderately Severe	Severe to Extreme	Total
.33	.71	.93	
567	1356	644	2588
5454	2773	1256	10479
6021	4129	1900	13067
553	1734	869	3174
4561	2516	1092	8848
5114	4250	1961	12022
11135	8379	3861	25089

5 x 16 matrix $\underset{\sim}{B}_1$ corresponding to H_1 has the same first 4 rows as $\underset{\sim}{B}$ and the fifth row is

$$(r_1,r_2,r_3,r_4,-r_1,-r_2,-r_3,-r_4,-r_1,-r_2,-r_3,-r_4,r_1,r_2,r_3,r_4),$$

that is, (.05,.33,.71,.93,-.05,-.33,-.71,-.93, ... ,.71,.93).

The vector $\underset{\sim}{\theta}_1$ equals (1,1,1,1,0)'.

The MDI estimates $x_1^*(ijk)$ of cell frequencies are

$x_1^*(111)$ = 19.95	$x_1^*(311)$ = 1359.58
$x_1^*(121)$ = 1004.56	$x_1^*(321)$ = 2759.92
$x_1^*(112)$ = 18.79	$x_1^*(312)$ = 1732.76
$x_1^*(122)$ = 671.88	$x_1^*(322)$ = 2529.14
$x_1^*(211)$ = 551.12	$x_1^*(411)$ = 657.35
$x_1^*(221)$ = 5469.99	$x_1^*(421)$ = 1244.54
$x_1^*(212)$ = 566.72	$x_1^*(412)$ = 855.73
$x_1^*(222)$ = 4543.51	$x_1^*(422)$ = 1103.47

The MDI statistic $2I(x_1^*:x)$ is 1.986 with 1 D.F. The value 2.02 was obtained by Bhapkar and Koch (1968) for their Wald-type statistic. The natural parameter tau associated with the "ridit" constraint is τ = 0.00841 and its variance is 0.000036. We note that $(0.00841)^2/0.000036 = 1.965$, a quadratic approximation to $2I(x_1^*:x)$.

The minimum modified chi-square estimates are

19.93	551.01	1359.74	657.32
1004.53	5470.01	2759.94	1244.52
18.78	566.60	1732.88	855.74
671.87	4543.54	2529.15	1103.43

and the value of the minimum modified chi-square is 1.979.

Now observe that H_1 is implied by the stronger hypothesis H_o given by (8) since the fifth row of $\underset{\sim}{B}_1$ can be expressed as a linear combination of rows of $\underset{\sim}{B}$. To see this let B(h) denote the h-th row of $\underset{\sim}{B}$ of (31), then

$$B_1(5) = r_1B(5) + r_2B(6) + r_3B(7) + r_4[B(1) - B(2) - B(3) + B(4) - B(5) - B(6) - B(7)].$$

Hence we can analyze the information $2I(x^*:x)$ as follows:

Analysis of Information

Component due to	Information	D.F.	Tabular Chi-square (5%)
H_0	$2I(x^*:x) = 19.703$	3	7.815
H_1	$2I(x^*:x_1^*) = 17.717$	2	5.991
	$2I(x_1^*:x) = 1.986$	1	3.841

We see that the data do not provide statistically significant evidence against the hypothesis H_1 of no second-order interaction with respect to average "ridits." In other words, this hypothesis does explain the departure from the hypothesis (8) of no linear second-order interaction.

Further analysis of these data can be done in two ways; in terms of "ridit" values and in terms of the non-quantitative contrasts among p(ijk) given by the last three rows of the matrix $\underset{\sim}{B}$ of (31).

A. "Ridits"

Note that the data are not consistent with the hypothesis of no linear second-order interaction ($2I(x^*:x) = 19.703$, 3 D.F.), while they can be regarded as consistent with the hypothesis H_1 of equality of differences of means of the "ridit" values (r_1,r_2,r_3,r_4) of the four distributions ($2I(x_1^*:x) = 1.986$, 1 D.F.). The remaining two degrees of freedom can be associated respectively with the hypotheses of equality of second and third moments of the "ridit" values. The hypothesis H_2 of equality of differences of means and of second moments corresponds to a 6 x 16 matrix, $\underset{\sim}{B}_2$, say, which has the first five rows as in $\underset{\sim}{B}_1$, the sixth row is

$(r_1^2, r_2^2, r_3^2, r_4^2, -r_1^2, -r_2^2, -r_3^2, -r_4^2, -r_1^2, -r_2^2, -r_3^2, -r_4^2, r_1^2, r_2^2, r_3^2, r_4^2)$

and $\underset{\sim}{\theta}_2 = (1,1,1,1,0,0)'$.

Under H_2 the MDI statistic $2I(x_2^*:x)$ is 10.036. The difference 10.036 - 1.980 = 8.056 is the contribution due to the additional constraint in $\underset{\sim}{B}_2$ as compared to $\underset{\sim}{B}_1$, assignable to equality of second moments. Finally the difference 19.703 - 10.036 = 9.667 is the contribution due to equality of the third moments in addition to the equality of the first two moments. Since each of these differences is asymptotically a chi-square with one degree of freedom, we conclude that though there is no significant second-order linear interaction with respect to mean "ridits," there appears to be a significant contribution due to heterogeneity of the second and third moments of the four "ridit" distributions.

B. Non-quantitative Approach

A different line of analysis treats the response variable (accident severity) as a qualitative variable ignoring "ridit" values. In this case, since the overall hypothesis of no linear second-order interaction leads to a significant MDI statistic ($2I(x^*:x) = 19.703$, 3 D.F.) it may be of interest to examine which of the three constraints (given by the last three rows of the matrix $\underset{\sim}{B}$ in (31)) contribute significantly to the value of $2I(x^*:x)$. For this purpose, we set up several B-matrices omitting one or two rows from the last three rows of (31) each time. For example, the B-matrix without the seventh row corresponds to the (weaker) hypothesis

$$H_3: \quad \begin{aligned} p(111) - p(112) - p(121) + p(122) &= 0, \\ p(211) - p(212) - p(221) + p(222) &= 0. \end{aligned}$$

Implicit in H_3 is the third constraint

$$[p(311) + p(411)] - [p(312) + p(412)] - [p(321) + p(421)] + [p(322) + p(422)] = 0.$$

Hence H_3 tests no linear second-order interaction with respect to levels 1 and 2 combining levels 3 and 4 of the response. Note that under these weaker hypotheses the MDI statistics will give a value not larger than 19.703. The analysis is summarized in Table 6.

Omitting rows 5 and 6 of the matrix $\underset{\sim}{B}$ of (31) corresponds to the hypothesis of no linear second-order interaction in a 2 x 2 x 2 table with level 3, pooling all the remaining levels. This is the only hypothesis with which the data are consistent. Thus it appears that levels 1 and 2 of accident severity both jointly and separately account for a major (significant) contribution towards the presence of a linear second-order interaction.

Table 6 also indicates a way of reducing categories in a contingency table with the inherent qualities of the observed data least affected. Thus if the given 4 x 2 x 2 table is to be reduced to a 3 x 2 x 2 table, this should be done by combining levels 3 and 4. Similarly, if a 2 x 2 x 2 table is required as a partial summary of the 4 x 2 x 2 table one should examine all the possible ways of pooling the levels of the response variable and select the way in which the maximum contribution to the linear second-order interaction is retained. The possible ways are level

(1) against (2) + (3) + (4), (2) against (1) + (3) + (4),
(3) against (1) + (2) + (4), (1) + (2) against (3) + (4),
(1) + (3) against (2) + (4), and (1) + (4) against (2) + (3).

The MDI statistics corresponding to the first three combinations are given in Table 6 as the three entries 11.803, 9.750, and 0.032 respectively. To find the MDI statistics corresponding to the remaining three combinations one can add the last two rows of the B-matrices when rows 7, 6, 5, are omitted one at a time. This gives the MDI statistics as 3.517, 1.538, and 8.078 respectively. The largest of these MDI statistics is 11.803, showing that levels 2, 3, and 4 should be pooled and level 1 be retained in the 2 x 2 x 2 table.

The analysis above shows that levels 1 and 2 are the main contributors to the departure from the hypothesis of no linear second-order interaction.

TABLE 6

MDI Statistics Under Different B-matrices

Operation on rows of (31)	MDI statistics	D.F.
Delete (7)	18.385	2
Delete (6)	12.125	2
Delete (5)	13.188	2
Delete (6), (7)	11.803	1
Delete (5), (7)	9.750	1
Delete (5), (6)	0.032	1
Delete (7), add (5) and (6)	3.517	1
Delete (6), add (5) and (7)	1.538	1
Delete (5), add (6) and (7)	8.088	1

We shall reexamine the data of this section from the standpoint of a different model in Example 8 of Chapter 8.

Chapter 8

FURTHER APPLICATIONS

In this chapter we shall consider eight examples illustrating the general analysis described in Chapter 5. Some of these examples illustrate aspects of the ECP methodology, in addition to the examples in Chapter 7. There are also some examples illustrating the application of the k-sample techniques to ICP. One of the examples of ECP shows the application of the MDI estimation procedure to count data where the random variables are quantitative and hypotheses about moments are tested. The use of a constraint involving a single cell is also shown. In the application of the k-sample technique to ICP there are also illustrations of the use of a distribution, other than the uniform, as the initial distribution of the iteration, and the computation of the natural parameters (taus) in such cases. In particular it should be noted that in the application of the k-sample technique to ICP the C-matrix is derived from the B-matrix by the relations

$$\underset{\sim}{B} = \begin{bmatrix} \underset{\sim}{B}_1 \\ \underset{\sim}{B}_2 \end{bmatrix}, \quad \underset{\sim}{C} = \begin{bmatrix} \underset{\sim}{C}_1 \\ \underset{\sim}{C}_2 \end{bmatrix}, \quad \underset{\sim}{C}_1 = \underset{\sim}{B}_1 \underset{\sim}{W}^{-1}, \quad \underset{\sim}{C}_2 = \underset{\sim}{B}_2,$$

where for k-samples the B_1-matrix consists of the k natural or normalizing constraints. This is because for ICP the B_2-matrix defines constraints to be satisfied by the MDI estimated occurrences rather than the estimated probabilities as in ECP.

EXAMPLE 1

Gail's data. This example illustrates the procedure for getting MDI estimates under hypotheses about the underlying probabilities of two contingency tables and testing the null hypothesis. An analysis of information table is also given in this case, including a subhypothesis. Note the difference in the analysis of information from those for the fitting problems.

GAIL'S DATA

As an illustration of the k-sample approach to ECP consider the following two contingency tables (artificial data) considered by Gail (1974, 97).

TABLE 1

20	5	5	30		15	15	2	32
6	4	2	12		10	5	5	20
26	9	7	42		25	20	7	52
	a)					b)		

The problem of interest was whether the underlying probabilities for the two tables were such that the respective marginal probabilities of the two tables were the same. If so, could it be a consequence of the fact that the tables were homogeneous? Let us denote the observed values in the two tables as in Table 2.

TABLE 2

x(111)	x(112)	x(113)	x(11.)		x(211)	x(212)	x(213)	x(21.)
x(121)	x(122)	x(123)	x(12.)		x(221)	x(222)	x(223)	x(22.)
x(1.1)	x(1.2)	x(1.3)	N_1		x(2.1)	x(2.2)	x(2.3)	N_2
	a)					b)		

For the hypothesis H_1 that the respective marginal probabilities are the same the basic values for the k-sample approach follow.

	ω	$x(\omega)$	$\ln x(\omega)$
111	1	20	2.995732
112	2	5	1.609438
113	3	5	1.609438
121	4	6	1.791759
122	5	4	1.386294
123	6	2	0.693147
211	7	15	2.708050
212	8	15	2.708050
213	9	2	0.693147
221	10	10	2.302585
222	11	5	1.609438
223	12	5	1.609438

$N_1 = 42$
$N_2 = 52$
$N = 94$
$w_1 = 42/94 = 0.446808$
$w_2 = 52/94 = 0.553191$
$v_1 = 1/w_1 = 2.238094$
$v_2 = 1/w_2 = 1.807692$

The B-matrix for H_1 and the values of $\underset{\sim}{\theta}$ and $N\underset{\sim}{\theta}$ are given in Table 3.

$$\underset{\sim}{W}_1 = \begin{bmatrix} w_1 & 0 & 0 & 0 & 0 & 0 \\ 0 & w_1 & 0 & 0 & 0 & 0 \\ 0 & 0 & w_1 & 0 & 0 & 0 \\ 0 & 0 & 0 & w_1 & 0 & 0 \\ 0 & 0 & 0 & 0 & w_1 & 0 \\ 0 & 0 & 0 & 0 & 0 & w_1 \end{bmatrix} \qquad \underset{\sim}{W}_2 = \begin{bmatrix} w_2 & 0 & 0 & 0 & 0 & 0 \\ 0 & w_2 & 0 & 0 & 0 & 0 \\ 0 & 0 & w_2 & 0 & 0 & 0 \\ 0 & 0 & 0 & w_2 & 0 & 0 \\ 0 & 0 & 0 & 0 & w_2 & 0 \\ 0 & 0 & 0 & 0 & 0 & w_2 \end{bmatrix}$$

$$\underset{\sim}{W} = \begin{bmatrix} \underset{\sim}{W}_1 & \underset{\sim}{0} \\ \underset{\sim}{0} & \underset{\sim}{W}_2 \end{bmatrix}, \quad \underset{\sim}{C} = \underset{\sim}{B}\underset{\sim}{W}^{-1}, \quad \underset{\sim}{C} = \begin{bmatrix} \underset{\sim}{C}_1 \\ \underset{\sim}{C}_2 \end{bmatrix}, \quad \underset{\sim}{C}_1 \text{ is } 2 \times 12, \ \underset{\sim}{C}_2 \text{ is } 3 \times 12$$

The C-matrix is obtained by multiplying all the elements in the first 6 columns of the B-matrix by 2.238094 and all the elements in the last 6 columns of the B-matrix by 1.807692.

$$\underset{\sim}{C}\underset{\sim}{x} = N\underset{\sim}{\phi} = \begin{bmatrix} 94 \\ 94 \\ 9.296700 \\ 12.998163 \\ -16.010971 \end{bmatrix}, \qquad \underset{\sim}{S} = \underset{\sim}{C}\underset{\sim}{D}_{\underset{\sim}{x}}\underset{\sim}{C}' = \begin{bmatrix} \underset{\sim}{S}_{11} & \underset{\sim}{S}_{12} \\ \underset{\sim}{S}_{21} & \underset{\sim}{S}_{22} \end{bmatrix},$$

$$\underset{\sim}{S}_{22.1}^{-1} = (\underset{\sim}{S}_{22} - \underset{\sim}{S}_{21}\underset{\sim}{S}_{11}^{-1}\underset{\sim}{S}_{12})^{-1} = \begin{bmatrix} 0.012198 & -0.001927 & -0.001776 \\ -0.001927 & 0.022284 & 0.017520 \\ -0.001776 & 0.017520 & 0.027001 \end{bmatrix},$$

$$\underset{\sim}{d} = N\underset{\sim}{\theta}^* - N\hat{\underset{\sim}{\theta}} = \begin{bmatrix} -9.296700 \\ -12.998163 \\ 16.010971 \end{bmatrix}.$$

The minimum modified chi-square value, a quadratic approximation to $2I(x^*:x)$ is $X^2 = \underset{\sim}{d}'\underset{\sim}{S}_{22.1}^{-1}\underset{\sim}{d} = 4.512$, 3 D.F. After 3 iterations the values of the MDI estimates are

	ω	x*(ω)	ℓn x*(ω)
111	1	16.504	2.803630
112	2	6.466	1.866627
113	3	4.042	1.396620
121	4	6.320	1.843785
122	5	6.604	1.887611
123	6	2.064	0.724458
211	7	18.263	2.904873
212	8	12.705	2.541984
213	9	2.476	0.906704
221	10	9.996	2.302228
222	11	3.477	1.246192
223	12	5.083	1.625813

It is found that $2I(x^*:x) = 4.333$, 3 D.F.

We now proceed to test the hypothesis H_2 that the two contingency tables are homogeneous. The B-matrix, $\underset{\sim}{\theta}$, and $N\underset{\sim}{\theta}$ for H_2 are given in Table 4. Using the B-matrix of Table 4 we have

$$\underset{\sim}{C} = \underset{\sim}{B}\underset{\sim}{W}^{-1}, \quad \underset{\sim}{C} = \begin{bmatrix} \underset{\sim}{C}_1 \\ \underset{\sim}{C}_2 \end{bmatrix}, \quad \underset{\sim}{C}_1 \text{ is } 2 \times 12, \quad \underset{\sim}{C}_2 \text{ is } 5 \times 12,$$

$$\underset{\sim}{C}\underset{\sim}{x} = N\underset{\sim}{\phi} = \begin{bmatrix} 94 \\ 94 \\ 17.646500 \\ -15.924902 \\ 7.575089 \\ -4.648350 \\ -0.086081 \end{bmatrix} .$$

$\underset{\sim}{S}_{22.1}^{-1}$ is now a 5 x 5 matrix, we omit the detailed values and $X^2 = \underset{\sim}{d}'\underset{\sim}{S}_{22.1}^{-1}\underset{\sim}{d} = 9.300$, 5 D.F. After 3 iterations the values of the MDI estimates are

	ω	$x^*(\omega)$	$\ell n\ x^*(\omega)$
111	1	15.901	2.766356
112	2	8.559	2.146948
113	3	2.808	1.032321
121	4	7.419	2.004110
122	5	4.219	1.439503
123	6	3.095	1.129803
211	7	19.686	2.979930
212	8	10.596	2.360522
213	9	3.476	1.245895
221	10	9.186	2.217686
222	11	5.223	1.653077
223	12	3.832	1.343370

It is found that under H_2, $2I(x^*:x) = 9.008$ 5 D.F.

If we denote the MDI estimate under the marginal homogeneity hypothesis H_1 by x_M^* and under the homogeneity hypothesis H_2 by x_H^*, then we may summarize the results in Table 5.

TABLE 3

	111	112	113	121	122	123	211	212	213	221	222	223		
ω	1	2	3	4	5	6	7	8	9	10	11	12	$\tilde{\theta}$	$N\tilde{\theta}$
	1	1	1	1	1	1	0	0	0	0	0	0	1	94
	0	0	0	0	0	0	1	1	1	1	1	1	1	94
	1	1	1	0	0	0	-1	-1	-1	0	0	0	0	0
	1	0	0	1	0	0	-1	0	0	-1	0	0	0	0
	0	1	0	0	1	0	0	-1	0	0	-1	0	0	0

We infer that the tables are homogeneous; hence the marginals are also homogeneous.

Note that

$$2\ \Sigma x_H^* \ \ell n(x_H^*/x) = 2\ \Sigma x_H^* \ \ell n(x_H^*/x_M^*) + 2\ \Sigma x_H^* \ \ell n(x_M^*/x).$$

But x_H^* also satisfies the constraints for x_M^* (homogeneity implies marginal homogeneity) hence

$$2\ \Sigma x_H^* \ \ell n(x_M^*/x) = 2\ \Sigma x_M^* \ \ell n(x_M^*/x)$$

and thus the analysis as in Table 5.

The statistics given by Gail (1974) are the same as the quadratic approximation X^2 values given above.

TABLE 4

ω	1	2	3	4	5	6	7	8	9	10	11	12	$\tilde{\theta}$	$N\tilde{\theta}$
	1	1	1	1	1	1	0	0	0	0	0	0	1	94
	0	0	0	0	0	0	1	1	1	1	1	1	1	94
	1	0	0	0	0	0	-1	0	0	0	0	0	0	0
	0	1	0	0	0	0	0	-1	0	0	0	0	0	0
	0	0	1	0	0	0	0	0	-1	0	0	0	0	0
	0	0	0	1	0	0	0	0	0	-1	0	0	0	0
	0	0	0	0	1	0	0	0	0	0	-1	0	0	0

TABLE 5

Analysis of Information

Component due to	Information	D.F.
H_2	$2I(x^*_H:x) = 9.008$	5
H_1	$2I(x^*_H:x^*_M) = 4.675$	2
	$2I(x^*_M:x) = 4.333$	3

EXAMPLE 2

Discrete distributions. This example illustrates the application of the k-sample procedure to test hypotheses about the means and variances of two discrete distributions, not in the form of contingency tables. An analysis of information table is given.

DISCRETE DISTRIBUTIONS

To illustrate the application of the k-sample procedure to quantitative variables in one-way contingency tables (discrete distributions) we consider the following example from Gokhale (1973), testing hypotheses about the means and variances. In this example $k = 2$, $\Omega_1 = (-2,-1,0,1,2)$, $\Omega_2 = (-1.5,1.5)$. The observed frequencies of the 5 cells of Ω_1 are respectively 6,18,9,24,3 and the observed frequencies of the 2 cells of Ω_2 are 72,48.

Let H_1 be the hypothesis that the population means in Ω_1 and Ω_2 are equal, and H_2 be the hypothesis that the population means <u>and</u> variances in Ω_1 and Ω_2 are equal. Note that H_2 is equivalent to the hypothesis that the first two population moments in Ω_1 and Ω_2 about the origin are respectively the same.

Under H_1 the constraints are $\underset{\sim}{B}\underset{\sim}{p} = \underset{\sim}{\theta}$ with

$$\underset{\sim}{B} = \begin{bmatrix} 1 & 1 & 1 & 1 & 1 & 0 & 0 \\ 0 & 0 & 0 & 0 & 0 & 1 & 1 \\ -2 & -1 & 0 & 1 & 2 & 1.5 & -1.5 \end{bmatrix}, \quad \underset{\sim}{\theta} = \begin{bmatrix} 1 \\ 1 \\ 0 \end{bmatrix}.$$

Note that the last row of the matrix $\tilde{B}$ derives from mean (Ω_1) - mean(Ω_2) = $-2p_1(-2) - 1p_1(-1) + 0p_1(0) + 1p_1(1) + 2p_1(2) - (-1.5p_2(-1.5) + 1.5p_2(1.5))$. Since $N_1 = 60$, $N_2 = 120$, $N = 180$, $w_1 = 1/3$, $w_2 = 2/3$,

$$\tilde{W}_1 = \begin{bmatrix} 1/3 & 0 & 0 & 0 & 0 \\ 0 & 1/3 & 0 & 0 & 0 \\ 0 & 0 & 1/3 & 0 & 0 \\ 0 & 0 & 0 & 1/3 & 0 \\ 0 & 0 & 0 & 0 & 1/3 \end{bmatrix} \quad \tilde{W}_2 = \begin{bmatrix} 2/3 & 0 \\ 0 & 2/3 \end{bmatrix},$$

$$\tilde{W} = \begin{bmatrix} \tilde{W}_1 & \tilde{0} \\ \tilde{0} & \tilde{W}_2 \end{bmatrix}, \quad \tilde{C} = \tilde{B}\tilde{W}^{-1} = \begin{bmatrix} 3 & 3 & 3 & 3 & 3 & 0 & 0 \\ 0 & 0 & 0 & 0 & 0 & 1.5 & 1.5 \\ -6 & -3 & 0 & 3 & 6 & 2.25 & -2.25 \end{bmatrix},$$

$$\tilde{x} = \begin{matrix} x(1) = 6 \\ x(2) = 18 \\ x(3) = 9 \\ x(4) = 24 \\ x(5) = 3 \\ x(6) = 72 \\ x(7) = 48 \end{matrix}, \qquad \begin{matrix} \ln x(1) = 1.791759 \\ \ln x(2) = 2.890371 \\ \ln x(3) = 2.197225 \\ \ln x(4) = 3.178054 \\ \ln x(5) = 1.098612 \\ \ln x(6) = 4.276666 \\ \ln x(7) = 3.871201 \end{matrix},$$

$$N\tilde{\theta} = \begin{bmatrix} 180 \\ 180 \\ 0 \end{bmatrix}, \qquad N\tilde{\phi} = \tilde{C}\tilde{x} = \begin{bmatrix} 180 \\ 180 \\ 54 \end{bmatrix},$$

$$\tilde{S} = \tilde{C}\tilde{D}_x\tilde{C}' = \begin{bmatrix} 540 & 0 & 0 \\ 0 & 270 & 81 \\ 0 & 81 & 1309.5 \end{bmatrix}, \qquad \tilde{S}_{22.1} = 1285.199951,$$

$$S_{22.1}^{-1} = 0.000778, \quad \tilde{\Delta} = \begin{bmatrix} 0 \\ 0 \\ -54 \end{bmatrix}, \quad d = -54,$$

$$X^2 = (-54)^2(.000778) = 2.269, \text{ 1 D.F.}$$

After two iterations the MDI estimates are

$x^*(1) = 7.618$	$\ln x^*(1) = 2.030505$	$-2 \times 7.618 = -15.236$
$x^*(2) = 20.180$	$\ln x^*(2) = 3.004673$	$-1 \times 20.180 = -20.180$
$x^*(3) = 8.909$	$\ln x^*(3) = 2.187082$	$0 \times 8.909 = 0$
$x^*(4) = 20.978$	$\ln x^*(4) = 3.043467$	$1 \times 20.978 = 20.978$
$x^*(5) = 2.315$	$\ln x^*(5) = 0.839582$	$2 \times 2.315 = 4.630$
$x^*(6) = 66.538$	$\ln x^*(6) = 4.197771$	$-1.5 \times 66.538 = -99.807$
$x^*(7) = 53.462$	$\ln x^*(7) = 3.978971$	$-1.5 \times 53.462 = 80.193$

$$2I(x^*:x) = 2.248, \text{ 1 D.F.}$$

$$(-15.236 - 20.180 + 0 + 20.978 + 4.630)/60 = -0.1635$$

$$(-99.807 + 80.193)/120 = -0.1635.$$

Under H_2 the restraints are $\underset{\sim}{B}\underset{\sim}{p} = \underset{\sim}{\theta}$ with

$$\underset{\sim}{B} = \begin{bmatrix} 1 & 1 & 1 & 1 & 1 & 0 & 0 \\ 0 & 0 & 0 & 0 & 0 & 1 & 1 \\ -2 & -1 & 0 & 1 & 2 & 1.5 & -1.5 \\ 4 & 1 & 0 & 1 & 4 & -2.25 & -2.25 \end{bmatrix}, \qquad \underset{\sim}{\theta} = \begin{bmatrix} 1 \\ 1 \\ 0 \\ 0 \end{bmatrix}.$$

Note that the last row of the matrix derives from

$$(-2)^2 p_1(-2) + (-1)^2 p_1(-1) + 0^2 p_1(0) + 1^2 p_1(1) + 2^2 p_1(2) -$$
$$((-1.5)^2 p_2(-1.5) + (1.5)^2 p_2(1.5)).$$

$$\underset{\sim}{C} = \underset{\sim}{B}\underset{\sim}{W}^{-1} = \begin{bmatrix} 3 & 3 & 3 & 3 & 3 & 0 & 0 \\ 0 & 0 & 0 & 0 & 0 & 1.5 & 1.5 \\ -6 & -3 & 0 & 3 & 6 & 2.25 & -2.25 \\ 12 & 3 & 0 & 3 & 12 & -3.375 & -3.375 \end{bmatrix},$$

$$N\underset{\sim}{\theta} = \begin{bmatrix} 180 \\ 180 \\ 0 \\ 0 \end{bmatrix}, \qquad \underset{\sim}{C}\underset{\sim}{x} = N\underset{\sim}{\phi} = \begin{bmatrix} 180 \\ 180 \\ 54 \\ -171 \end{bmatrix},$$

$$\underset{\sim}{S} = \underset{\sim}{C}\underset{\sim}{D}_x\underset{\sim}{C}' = \begin{bmatrix} 540 & 0 & 0 & 702 \\ 0 & 270 & 81 & -607.5 \\ 0 & 81 & 1309.5 & -344.25 \\ 702 & -607.5 & -344.25 & 3040.875 \end{bmatrix},$$

$$\underset{\sim}{S}_{22.1} = \begin{bmatrix} 1285.199951 & -162.0 \\ -162.0 & 761.399902 \end{bmatrix},$$

$$\underset{\sim}{S}_{22.1}^{-1} = \begin{bmatrix} .000800 & .000170 \\ .000170 & .001350 \end{bmatrix}, \quad \underset{\sim}{\Delta} = \begin{bmatrix} 0 \\ 0 \\ -54 \\ 171 \end{bmatrix}, \quad \underset{\sim}{d} = \begin{bmatrix} -54 \\ 171 \end{bmatrix},$$

$$X^2 = \underset{\sim}{d}'\underset{\sim}{S}_{22.1}^{-1}\underset{\sim}{d} = 38.652, \text{ 2 D.F.}$$

After four iterations the MDI estimates are

$x^*(1) = 18.134$ $\qquad \ell n\ x^*(1) = 2.897783$

$x^*(2) = 13.081$ $\qquad \ell n\ x^*(2) = 2.571174$

$x^*(3) = 4.000$ $\qquad \ell n\ x^*(3) = 1.386189$

$x^*(4) = 16.586$ $\qquad \ell n\ x^*(4) = 2.808560$

$x^*(5) = 8.199$ $\qquad \ell n\ x^*(5) = 2.104045$

$x^*(6) = 70.910$ $\qquad \ell n\ x^*(6) = 4.261405$

$x^*(7) = 49.090$ $\qquad \ell n\ x^*(7) = 3.893661$

$$2I(x^*:x) = 29.546,\ 2\ \text{D.F.}$$

We leave it as an exercise for the reader to verify that the MDI estimates satisfy the moment constraints.

We summarize in the analysis of information table.

Analysis of Information

Component due to	Information	D.F.
H_2	$2I(x_2^*:x) = 29.546$	2
H_2-H_1 (Effect)	$2I(x_2^*:x_1^*) = 27.298$	1
H_1	$2I(x_1^*:x) = 2.248$	1

We reject the hypothesis H_2 but accept the hypothesis H_1. The effect of the differences in the variances is significant.

EXAMPLE 3

Marginal homogeneity of an r x r contingency table. This example illustrates the application of the single sample procedure to a set of data previously studied using a different algorithm. It also serves as an introduction to the next example. This example shows that the MDI estimate retains properties of the original observations not involved in the null hypothesis. For applications of the notion of marginal homogeneity to higher order contingency tables see Kullback (1971a, 1971b). The latter paper includes an example of the quadratic approximation to $2I(x^*:x)$. The analysis in the example follows the discussion and notation of Chapter 5, Section III.

MARGINAL HOMOGENEITY OF AN $r \times r$ CONTINGENCY TABLE

In Ireland, Ku and Kullback (1969) the principle of MDI estimation was applied to obtain MDI estimates of the cell frequencies of an r x r contingency table under hypotheses of symmetry and marginal homogeneity using a special iterative algorithm. The procedures were illustrated with data from case-records of the eye-testing of employees in Royal Ordnance factories analysed by Stuart (1955).

TABLE 1

7477 Women Aged 30-39; Unaided Distance Vision

x(ij)

Left Eye Right Eye	Highest Grade	Second Grade	Third Grade	Lowest Grade	Total
Highest Grade	1520	266	124	66	1976
Second Grade	234	1512	432	78	2256
Third Grade	117	362	1772	205	2456
Lowest Grade	36	82	179	492	789
	1907	2222	2507	841	7477

In this example we use the single sample iterative algorithm to derive the MDI estimates x^* as well as $\tilde{x}$ the minimum modified chi-square estimates. These values are given in Table 3. We relate the results to values given by Stuart (1955) and Bhapkar (1966). The basic data will also be used in the next example to illustrate the k-sample algorithm applied to incomplete data.

We remind the reader that the graphic form of the loglinear representation with design matrix $\underset{\sim}{C}'$ represents

$$\ln (x^*(\omega)/x(\omega)) = L + \tau_1 c_1(\omega) + \tau_2 c_2(\omega) + \tau_3 c_3(\omega).$$

It is found that $L = 0.000805$, $\tau_1 = -0.159043$, $\tau_2 = -0.105379$, $\tau_3 = -0.050000$, where the covariance matrix of the natural parameters τ_1, τ_2, τ_3 is

$$\begin{bmatrix} 0.002500 & 0.001570 & 0.001304 \\ 0.001570 & 0.001976 & 0.001372 \\ 0.001304 & 0.001372 & 0.001695 \end{bmatrix}$$

LOGLINEAR REPRESENTATION

ij	ω	L	τ_1	τ_2	τ_3
11	1	1	0	0	0
12	2	1	1	-1	0
13	3	1	1	0	-1
14	4	1	1	0	0
21	5	1	-1	1	0
22	6	1	0	0	0
23	7	1	0	1	-1
24	8	1	0	1	0
31	9	1	-1	0	1
32	10	1	0	-1	1
33	11	1	0	0	0
34	12	1	0	0	1
41	13	1	-1	0	0
42	14	1	0	-1	0
43	15	1	0	0	-1
44	16	1	0	0	0

$$N\underset{\sim}{\theta}' = (7477,0,0,0)$$

From the loglinear representation of the MDI estimate we see that certain associations in the original table are the same as in the estimated table, thus

$$\ln x^*(ii)x^*(jj)/x^*(ij)x^*(ji) = \ln x(ii)x(jj)/x(ij)x(ji),$$

$$\ln x^*(ij)x^*(44)/x^*(i4)x^*(4j) = \ln x(ij)x(44)/x(i4)x(4j).$$

Bhapkar's test statistic $\underset{\sim}{d}'\underset{\sim}{S}_{22.1}^{-1}\underset{\sim}{d}$, is the minimum modified chi-square and he gave $X_B^2 = 11.976$ with 3 D.F. He did not give the minimum modified chi-square estimates. We obtained $X^2 = 11.9757$. Stuart gave no estimates either and he used as his statistic $X_S^2 = \underset{\sim}{d}'\underset{\sim}{S}_{22}^{-1}\underset{\sim}{d} = 11.957$. Stuart estimated the covariance matrix of the $\underset{\sim}{d}$'s under the null hypothesis. From Table 2 we see that $\underset{\sim}{S}_{22}$ and $\underset{\sim}{S}_{22.1}$ are not very much different in this case. We leave it as an exercise for the reader to show that (Ireland, Ku and Kullback, 1969) $X_B^2 = X_S^2/(1 - X_S^2/n)$.

Based on the values $\chi_S^2 = 11.957$, $\chi_B^2 = 11.976$ with 3 D.F., Stuart, and also Bhapkar, rejected the null hypothesis of marginal homogeneity. We find that $2I(x^*:x) = 12.017$, 3 D.F. and reject the null hypothesis of homogeneity.

TABLE 2

$$\underset{\sim}{S} = \underset{\sim}{C}\underset{\sim}{D}_x\underset{\sim}{C}'$$

$$\begin{bmatrix} 7477 & 69 & 34 & -51 \\ 69 & 843 & -500 & -241 \\ 34 & -500 & 1454 & -794 \\ -51 & -241 & -794 & 1419 \end{bmatrix}$$

$$\underset{\sim}{S}_{22.1} = \underset{\sim}{S}_{22} - \underset{\sim}{S}_{21}\underset{\sim}{S}_{11}^{-1}\underset{\sim}{S}_{12}$$

$$\begin{bmatrix} 842.363037 & -500.313721 & -240.529343 \\ -500.313721 & 1453.845215 & -793.768066 \\ -240.529343 & -793.768066 & 1418.652100 \end{bmatrix}$$

TABLE 3

ω	$x(\omega)$	$x^*(\omega)$	$\tilde{x}(\omega)$
1	1520	1521.222	1522.434
2	266	252.304	252.212
3	124	111.279	110.707
4	66	56.341	55.642
5	234	247.010	246.879
6	1512	1513.217	1514.422
7	432	409.056	408.775
8	78	70.255	69.927
9	117	130.585	129.917
10	362	382.920	382.621
11	1772	1773.425	1774.838
12	205	195.159	195.133
13	36	42.240	41.765
14	82	91.186	90.749
15	179	188.329	188.189
16	492	492.396	492.788

EXAMPLE 4

Several samples, incomplete data.

This example uses the complete contingency table of the preceding example and row and column marginals only of additional samples. The example illustrates the application of ECP to samples which may include fragmentary data.

SEVERAL SAMPLES, INCOMPLETE DATA

We shall illustrate the k-sample algorithm of testing several samples with some incomplete data in terms of a specific sample. In Table 1 the 7477 observations in the 4 x 4 contingency table are Stuart's data, which we have already examined under the null hypothesis of marginal homogeneity.

"The remaining 1100 observations are artificial data for 600 women for whom only left eye vision was reported and 500 women for whom only right eye vision was reported. It will be presumed that the incomplete data for women with vision classified only for one eye arose in a completely random manner which was statistically independent of the true classification of their vision with respect to both eyes. This assumption allows us to say that the marginal probabilities pertaining to left eye vision and right eye vision for women classified on both eyes are the same parameters as the probabilities pertaining to left eye vision for women only for the left eye and to right eye vision for women classified only for the right eye respectively" (Koch, Imrey and Reinfurt, 1972, 665-6).

The results for the k-sample algorithm are summarized in Table 2, in which we also give the values derived by Koch, Imrey and Reinfurt (1972) by their approach. Table 3 gives the value of the B-matrix and $N\underset{\sim}{\theta}$.

In view of the small values of the test statistics with 6 D.F. we accept the null hypothesis of the homogeneity of the 3 sets of data with respect to the underlying population.

Using the MDI estimates of the entries in the cells of the complete contingency table as "improved" values over the original observations we repeat the test for the null hypothesis of marginal homogeneity. The resulting values are summarized in Table 2. There is no change in the inference that the data show no evidence of marginal homogeneity.

Table 4 gives the graphic version of the loglinear representations which are

$$\ln(x^*(1)/x(1)) = v_1L_1 + v_1\tau_1 + v_1\tau_4$$

$$\ln(x^*(2)/x(2)) = v_1L_1 + v_1\tau_1 + v_1\tau_5$$

$$\vdots \qquad \vdots$$

$$\ln(x^*(16)/x(16)) = v_1L_1$$

$$\ln(x^*(17)/x(17)) = v_2L_2 - v_2\tau_1$$

$$\vdots \qquad \vdots$$

$$\ln(x^*(20)/x(20)) = v_2L_2$$

$$\ln(x^*(21)/x(21)) = v_3L_3 - v_3\tau_3$$

$$\vdots \qquad \vdots$$

$$\ln(x^*(24)/x(24)) = v_3L_3$$

The numerical values of the inverse weights and natural parameters in the loglinear representations are $v_1 = 1/w_1 = 1.147118$, $v_2 = 1/w_2 = 17.153992$, $v_3 = 1/w_3 = 14.294999$

$$\tau_1 = 0.005890, \ \tau_2 = 0.002315, \ \tau_3 = 0.001283$$

$$\tau_4 = 0.010207, \ \tau_5 = 0.008115, \ \tau_6 = 0.007189.$$

We see that $\ln(x^*(1)/x(1)) - \ln(x^*(4)/x(4)) = v_1\tau_4 = \ln(x^*(5)/x(5)) - \ln(x^*(8)/x(8))$ or $\ln x^*(1)x^*(8)/x^*(4)x^*(5) = \ln x(1)x(8)/x(4)x(5)$. We leave it as an exercise for the reader to determine other associations in the MDI estimate that are inherited from the original data.

TABLE 1

UNAIDED DISTANCE VISION; 8577 WOMEN AGED 30-39

Right Eye	Left eye Highest Grade (1)	Second Grade (2)	Third Grade (3)	Lowest Grade (4)	Sub-Total	Right Only	Total
Highest Grade (1)	1520	266	124	66	1976	140	2116
Second Grade (2)	234	1512	432	78	2256	150	2406
Third Grade (3)	117	362	1772	205	2456	160	2616
Lowest Grade (4)	36	82	179	492	789	50	839
Sub-Total	1907	2222	2507	841	7477	500	7977
Left Only	160	180	200	60	600	*	*
Total	2067	2402	2707	901	8077	*	8577

See Koch, G.G., Imrey, P.B., and Reinfurt, D.W. (1972), in particular page 665.

TABLE 2

ij	ω	x(ω)	x*(ω)	$\tilde{x}(\omega)$	$\hat{x}(\omega)^a$	$\tilde{\tilde{x}}(\omega)^b$	$x^{**}(\omega)^c$
11	1	1520	1530.227	1530.155	1529.495	1532.573	1531.372
12	2	266	267.148	267.151	266.331	253.107	253.216
13	3	124	124.403	124.405	123.670	110.966	111.552
14	4	66	65.671	65.643	65.573	55.529	56.202
21	5	234	234.664	234.676	235.301	247.726	247.955
22	6	1512	1512.657	1512.810	1512.672	1515.085	1513.898
23	7	432	431.729	431.773	430.600	408.454	408.742
24	8	78	77.311	77.282	77.387	69.555	69.857
31	9	117	117.190	117.195	117.838	130.215	130.894
32	10	362	361.721	361.751	362.784	382.343	382.648
33	11	1772	1768.752	1768.905	1769.357	1771.597	1770.209
34	12	205	202.944	202.863	203.748	193.837	193.836
41	13	36	36.006	36.007	36.114	41.662	42.121
42	14	82	81.818	81.822	81.798	90.284	90.690
43	15	179	178.413	178.422	177.878	186.975	187.084
44	16	492	486.360	486.142	486.528	487.092	486.710
1.	17	140	132.904	132.898	132.745		
2.	18	150	150.887	150.899	150.860		
3.	19	160	163.876	163.884	164.085		
4.	20	50	52.333	52.320	52.310		
.1	21	160	153.919	153.915	153.966		
.2	22	180	178.415	178.430	178.434		
.3	23	200	200.880	200.897	200.736		
.4	24	60	66.787	66.759	66.864		
			2I(x*:x) = 1.771 6 D.F.	X^2 = 1.764 6 D.F.	X^2 = 2.33 6 D.F.	X^2 = 11.741 3 D.F.	2I(x**:x*) = 11.730 3 D.F.

a) See Koch, Imrey and Reinfurt (1972, 669)

b),c) Using "improved" estimate to test marginal homogeneity

b) is minimum modified chi-square estimate and c) is MDI estimate

TABLE 3

	ω	1	2	3	4	5	6	7	8	9	10	11	12	13	14	15	16	17	18	19	20	21	22	23	24
		1	1	1	1	1	1	1	1	1	1	1	1	1	1	1	1	0	0	0	0	0	0	0	0
		0	0	0	0	0	0	0	0	0	0	0	0	0	0	0	0	1	1	1	1	0	0	0	0
		0	0	0	0	0	0	0	0	0	0	0	0	0	0	0	0	0	0	0	0	1	1	1	1
$\tilde{B}$ =		1	1	1	1	0	0	0	0	0	0	0	0	0	0	0	0	-1	0	0	0	0	0	0	0
		0	0	0	0	1	1	1	1	0	0	0	0	0	0	0	0	0	-1	0	0	0	0	0	0
		0	0	0	0	0	0	0	0	1	1	1	1	0	0	0	0	0	0	-1	0	0	0	0	0
		1	0	0	0	1	0	0	0	1	0	0	0	1	0	0	0	0	0	0	0	-1	0	0	0
		0	1	0	0	0	1	0	0	0	1	0	0	0	1	0	0	0	0	0	0	0	-1	0	0
		0	0	1	0	0	0	1	0	0	0	1	0	0	0	1	0	0	0	0	0	0	0	-1	0

$N\tilde{\theta}' = (8577\ 8577\ 8577\ 0\ 0\ 0\ 0\ 0\ 0)$

TABLE 4

Loglinear Representation

ij	ω	L_1	L_2	L_3	τ_1	τ_2	τ_3	τ_4	τ_5	τ_6
11	1	v_1			v_1			v_1		
12	2	v_1			v_1				v_1	
13	3	v_1			v_1					v_1
14	4	v_1			v_1					
21	5	v_1				v_1		v_1		
22	6	v_1				v_1			v_1	
23	7	v_1				v_1				v_1
24	8	v_1				v_1				
31	9	v_1					v_1	v_1		
32	10	v_1					v_1		v_1	
33	11	v_1					v_1			v_1
34	12	v_1					v_1			
41	13	v_1						v_1		
42	14	v_1							v_1	
43	15	v_1								v_1
44	16	v_1								
1.	17		v_2		$-v_2$					
2.	18		v_2			$-v_2$				
3.	19		v_2				$-v_2$			
4.	20		v_2							
.1	21			v_3			$-v_3$			
.2	22			v_3				$-v_3$		
.3	23			v_3					$-v_3$	
.4	24			v_3						

EXAMPLE 5

Specified loglinear representation.

In this example the problem specifies the form of the loglinear representation and consequently the design matrix. The general linear hypothesis approach is necessary.

SPECIFIED LOGLINEAR REPRESENTATION

Gokhale (1972) formulated a problem for a 2 x 2 x 3 three-way contingency table of fitting a model such that the loglinear representation is of the form

$$\begin{aligned}\ell n(x^*(ijk)/n\pi) = {} & L + (i-1)\tau^{i} + (j-1)\tau^{j} + (k-1)\tau^{k} \\ & + (i-1)(j-1)\tau^{ij} + (i-1)(k-1)\tau^{ik} \\ & + (j-1)(k-1)\tau^{jk} + (i-1)(j-1)(k-1)\tau^{ijk}.\end{aligned}$$

This implies that the graphic version of the loglinear representation is as given in Table 1.

TABLE 1

		1	2	3	4	5	6	7	8
ω	ijk	L	τ^{i}	τ^{j}	τ^{k}	τ^{ij}	τ^{ik}	τ^{jk}	τ^{ijk}
1	111	1	0	0	0	0	0	0	0
2	112	1	0	0	1	0	0	0	0
3	113	1	0	0	2	0	0	0	0
4	121	1	0	1	0	0	0	0	0
5	122	1	0	1	1	0	0	1	0
6	123	1	0	1	2	0	0	2	0
7	211	1	1	0	0	0	0	0	0
8	212	1	1	0	1	0	1	0	0
9	213	1	1	0	2	0	2	0	0
10	221	1	1	1	0	1	0	0	0
11	222	1	1	1	1	1	1	1	1
12	223	1	1	1	2	1	2	2	2

The observed values (fictitious) are

ijk	x(ijk)	ijk	x(ijk)
111	58	211	75
112	49	212	58
113	33	213	45
121	11	221	19
122	14	222	17
123	18	223	22

We shall derive the MDI estimate using the k-sample algorithm (of course here k = 1) with the uniform distribution as the initial distribution. In this case the C-matrix is the transpose of the T-matrix in Table 1 and is given again for convenience in Table 2.

TABLE 2

i	1	1	1	1	1	1	2	2	2	2	2	2
j	1	1	1	2	2	2	1	1	1	2	2	2
k	1	2	3	1	2	3	1	2	3	1	2	3
ω	1	2	3	4	5	6	7	8	9	10	11	12
	1	1	1	1	1	1	1	1	1	1	1	1
	0	0	0	0	0	0	1	1	1	1	1	1
	0	0	0	1	1	1	0	0	0	1	1	1
	0	1	2	0	1	2	0	1	2	0	1	2
	0	0	0	0	0	0	0	0	0	1	1	1
	0	0	0	0	0	0	0	1	2	0	1	2
	0	0	0	0	1	2	0	0	0	0	1	2
	0	0	0	0	0	0	0	0	0	0	1	2

The moment constraints are

$$\underset{\sim}{T}'\underset{\sim}{x}^* = \underset{\sim\sim}{Cx}^* = \underset{\sim}{T}'\underset{\sim}{x} = \underset{\sim\sim}{Cx} = \begin{bmatrix} 419 \\ 236 \\ 101 \\ 374 \\ 58 \\ 209 \\ 111 \\ 61 \end{bmatrix}$$

The MDI estimated values are

ijk	x*(ijk)	ijk	x*(ijk)
111	59.73	211	74.97
112	45.54	212	58.06
113	34.73	213	44.97
121	10.98	221	17.85
122	14.05	222	19.29
123	17.98	223	20.85

The value of $2I(x:x^*)$ is 0.814 with 4 D.F.

EXAMPLE 6

Four point bioassay-fit of logistic function.

This example illustrates the application of the k-sample procedure to fitting data based on constraints using the observed values, that is, ICP. A similar procedure is also used in some of the following examples. This example also illustrates the computation of the natural parameters (taus) when the initial distribution of the iteration is not the uniform distribution.

FOUR POINT BIOASSAY-FIT OF LOGISTIC FUNCTION

The k-sample algorithm for MDI estimation may also be applied in ICP. We shall illustrate the procedure with a set of artificial data given by Berkson (1972, 446).

The problem as formulated by Berkson is, "Suppose P_i in relation to a quantity x_i is given by the logistic function

$$P_i = [1 + \exp(-(\alpha + \beta x_i))]^{-1}.$$

In Table 1 are given 4 observations, as for a 'bioassay' experiment, the data having been fabricated for expository purposes."

TABLE 1

x	Total	Observed deaths
0	10	1
1	10	6
2	10	3
3	10	8
Total	40	18

We remark that $Q_i = 1 - P_i = \exp(-(\alpha + \beta x_i))P_i$ so that the log-odds or logistic is $\ell n\, P_i/Q_i = \alpha + \beta x_i$.

We reformulate the data first as the 4 x 2 contingency Table 2, with entries x(ij), i = 1, ... , 4, j = 1,2

TABLE 2

		j = 1 Deaths	j = 2 Alive	
i	1	1	9	10
	2	6	4	10
	3	3	7	10
	4	8	2	10
		18	22	40

The graphic version of the loglinear representation of the MDI estimate is shown in Table 3.

TABLE 3

ω	ij	L_1	L_2	L_3	L_4	τ_1	τ_2
1	11	1				1	0
2	12	1				0	0
3	21		1			1	1
4	22		1			0	0
5	31			1		1	2
6	32			1		0	0
7	41				1	1	3
8	42				1	0	0

For the ICP fitting procedure we note that Table 3 implies the following relations.

$$x^*(i1) + x^*(i2) = x(i1) + x(i2),\ i = 1,2,3,4\ ,$$

$$x^*(11) + x^*(21) + x^*(31) + x^*(41) = x(11) + x(21) + x(31) + x(41),$$

$$x^*(21) + 2x^*(31) + 3x^*(41) = x(21) + 2x(31) + 3x(41),$$

$$\ell n\ (x^*(11)/x^*(12)) = \tau_1,$$

$$\ell n\ (x^*(21)/x^*(22)) = \tau_1 + \tau_2,$$

$$\ell n\ (x^*(31)/x^*(32)) = \tau_1 + 2\tau_2,$$

$$\ell n\ (x^*(41)/x^*(42)) = \tau_1 + 3\tau_2.$$

For the k-sample algorithm this is a case of 4 samples, two observations per sample. The basic B-matrix is given in Table 4.

TABLE 4

	11	12	21	22	31	32	41	42
ω	1	2	3	4	5	6	7	8
	1	1	0	0	0	0	0	0
	0	0	1	1	0	0	0	0
	0	0	0	0	1	1	0	0
	0	0	0	0	0	0	1	1
	1	0	1	0	1	0	1	0
	0	0	1	0	2	0	3	0

In view of the relations given above between values of the x^*'s and the x's the last two rows of the B-matrix are constraints the MDI estimated values must satisfy. In this case the C-matrix is derived from the B-matrix by the relations

$$\underset{\sim}{B}' = (\underset{\sim}{B}_1', \underset{\sim}{B}_2'),\ \underset{\sim}{C}' = (\underset{\sim}{C}_1', \underset{\sim}{C}_2'),\ \underset{\sim}{C}_1 = \underset{\sim}{C}_1 \underset{\sim}{W}^{-1},\ \underset{\sim}{C}_2 = \underset{\sim}{B}_2,$$

where $\underset{\sim}{B}$ is 6 x 8, $\underset{\sim}{B}_1$ is 4 x 8, $\underset{\sim}{B}_2$ is 2 x 8 with similar dimensions for the C-matrix and its components.

We remark that instead of starting the iteration from the uniform distribution, the initial distribution used in the computation was $x_1^*(ij) = x(i.)x(.j)/N$ as calculated from Table 2. We comment that another computation using the uniform distribution $N\pi(ij) = 5$ as the initial distribution for the iteration, of course, yielded the same final MDI estimated values.

By computing the maximum likelihood estimates of α and β in his formulation, Berkson derived the estimates given in Table 5.

TABLE 5

Berkson's Estimate Maximum Likelihood

Deaths	Alive	
1.901431	8.098569	10.000000
3.445099	6.554901	10.000000
5.405505	4.594495	10.000000
7.247965	2.752035	10.000000
18.000000	22.000000	40.000000

Berkson gave a value $2I(x:x^*) = 5.985432$, 2 D.F. (in Berkson (1972, 447) the degrees of freedom are misprinted as 1).

The MDI estimates after 4 iterations are given in Table 6.

TABLE 6

MDI Estimate Four Iterations

Deaths	Alive	
1.901434	8.098566	10.000000
3.445101	6.554895	9.999996
5.405508	4.594491	9.999999
7.247968	2.752036	10.000004
18.000011	21.999988	39.999999

$2I(x:x^*) = 5.985401$, 2 D.F.

We also have the analysis of information

Analysis of Information

Component due to	Information	D.F.
$x(i.), x(.j)$	$2I(x:x_1^*) = 12.863$	3
$x(.j)$	$2I(x^*:x_1^*) = 6.878$	1
$x(21) + 2x(31) + 3x(41), x(i.)$	$2I(x:x^*) = 5.985$	2

From Table 3 we see that since x_1^* was the initial distribution

$$\ell n\ (x^*(1)/x^*(2)) = \ell n\ (x_1^*(1)/x_1^*(2)) + \tau_1 \text{ or}$$

$$-1.449079 = -0.200671 - 1.248407,$$

$$\ell n\ (x^*(3)/x^*(4)) = \ell n\ (x_1^*(3)/x_1^*(4)) + \tau_1 + \tau_2$$

$$\text{or } -0.643259 = -0.200671 -1.248407 + 0.805820,$$

$$\ell n\ (x^*(5)/x^*(6)) = \ell n\ (x_1^*(5)/x_1^*(6)) + \tau_1 + 2\tau_2$$

$$\text{or } 0.162560 = -0.200671 -1.248407 + 1.611640,$$

$$\ln(x^*(7)/x^*(8)) = \ln(x_1^*(7)/x_1^*(8)) + \tau_1 + 3\tau_2$$

$$\text{or} \quad 0.968380 = -0.200671$$
$$-1.248407 + 2.417460,$$

or

$$\ln(x^*(3)/x^*(4)) - \ln(x^*(1)/x^*(2)) = 0.805820,$$

$$\ln(x^*(5)/x^*(6)) - \ln(x^*(3)/x^*(4)) = 0.805819,$$

$$\ln(x^*(7)/x^*(8)) - \ln(x^*(5)/x^*(6)) = 0.805820.$$

Since the matrix $\underset{\sim}{S}_{22.1}^{-1}$ for the observed distribution is

$$\begin{bmatrix} 0.282828 & -0.121212 \\ -0.121212 & 0.080808 \end{bmatrix}$$

we have a quadratic approximation to $2I(x^*:x_1^*)$ given by $(9)^2(0.080808) = 6.545$. The chi-square approximation to $2I(x^*:x_1^*)$ is given by

$$\Sigma(x^*(ij) - x_1^*(ij))^2/x_1^*(ij)$$

$$= (10/(4.5)(5.5))((2.599)^2 + (1.055)^2 + (0.906)^2 + (2.748)^2)$$

$$= 6.562.$$

EXAMPLE 7

Respiratory data. This example deals with two three-way 9 x 2 x 2 contingency tables which are marginal tables of a higher order dimensional table, not available to us, listing data on respiratory symptoms among a group of British coal miners. We illustrate the use of OUTLIER to partition second-order interaction in a three-way contingency table. Also illustrated are multivariate logit analysis and the relations among the natural parameters implied by logit linearity. MDI estimates under the hypothesis of logit linearity are obtained. We also obtain the MDI estimates under the hypothesis that associations as measured by the logarithm of cross-product ratios are linearly dependent on age (Plackett, 1974, 95). Confidence intervals for the associated natural parameter (tau) are also given. This example covers a broad spectrum of the methodology and iterative algorithms.

RESPIRATORY DATA

This example deals with two three-way contingency tables arising from respiratory symptoms among the same group of British coal miners. The analyses progressively consider more complex hypotheses because of basic differences in certain properties of the two sets of data. Among other features the example illustrates a test of the hypothesis of no second-order interaction in a three-way contingency table, multivariate logit analysis, the partitioning of second-order interaction in a three-way contingency table, and the use of the k-sample technique for an ICP.

The analyses are based on the principle of MDI estimation, the associated loglinear representation and analysis of information tables as discussed in Chapters 3 and 5. The computational procedures for this example utilize the iterative algorithms described in Chapters 5 and 6. Since certain MDI estimates are constrained to satisfy linear relations based on observed values, they are maximum likelihood estimates and the associated MDI test statistics are log-likelihood ratio statistics.

In Grizzle (1971) a model developed by Grizzle, Starmer, and Koch (1969) is specialized to the case of fitting models to correlated logits. Grizzle (1971, 1060) says, "Unfortunately a test of the goodness-of-fit of the logit model to the joint response data has not been developed." For its methodological interest, we first consider the problem as presented by Grizzle (1971) but from the MDI estimation approach. Our results (maximum likelihood) are numerically in close agreement with those of Grizzle (BAN), but also include MDI estimates of the cell entries under the logit model and a test of the goodness-of-fit to the joint response data.

In Table 1 is given a 9 x 2 x 2 contingency table of coal-miners classified as smokers without radiological pneumoconiosis, between the ages of 20 and 64 years inclusive at the time of their examination, showing the occurrence of breathlessness and wheeze over nine age groupings. We denote the observed frequency in any cell by x(ijk) with

Variable		Index	1	2	3	4	...	9
Age Group	A	i	20-24	25-29	30-34	35-39	...	60-64
Breathlessness	B	j	yes	no				
Wheeze	W	k	yes	no				

These data are discussed and analysed from a different point of view by Ashford and Sowden (1970), Mantel and Brown (1973).

A loglinear representation of the observed values x(ijk) in Table 1 is given in columns 1-36 of Fig. 1. Fig. 1 is a graphic presentation of the design matrix of the complete loglinear representation

$$
\begin{aligned}
\ell n\, x(ijk)/n\pi(ijk) &= L + \tau_1^A T_1^A(ijk) + \ldots + \tau_8^A T_8^A(ijk) \\
&+ \tau_1^B T_1^B(ijk) + \tau_1^W T_1^W(ijk) + \tau_{11}^{AB} T_{11}^{AB}(ijk) + \ldots + \tau_{81}^{AB} T_{81}^{AB}(ijk) \\
&+ \tau_{11}^{AW} T_{11}^{AW}(ijk) + \ldots + \tau_{81}^{AW} T_{81}^{AW}(ijk) + \tau_{11}^{BW} T_{11}^{BW}(ijk) \\
&+ \tau_{111}^{ABW} T_{111}^{ABW}(ijk) + \ldots + \tau_{811}^{ABW} T_{811}^{ABW}(ijk),
\end{aligned} \tag{1}
$$

where $\pi(ijk) = 1/36$, n is the total number of observations, L is a normalizing factor (the negative of the logarithm of a moment generating function) and the T(ijk) are linearly independent indicator functions (explanatory variables) taking on the values given by the columns of Fig. 1 and whose mean values are the various marginals.

Since Grizzle (1971) is concerned with the marginal logits of breathlessness and wheeze, this means implicitly that one is concerned with the MDI estimate, or loglinear representation, obtained by fitting the marginals x(ij.) and x(i.k). Denote this estimate by $x_d^*(ijk)$, then its loglinear representation or design matrix is given by columns 1-27 of Fig. 1. It may be verified that x_d^* has the explicit form $x_d^*(ijk) = x(ij.)x(i.k)/x(i..)$ and consequently we have the marginal logits

$$\ell n \frac{x_d^*(i1k)}{x_d^*(i2k)} = \ell n \frac{x(i1.)x(i.k)x(i..)}{x(i..)x(i2.)x(i.k)} = \ell n \frac{x(i1.)}{x(i2.)} \quad \text{(breathlessness)}$$

$$\ell n \frac{x_d^*(ij1)}{x_d^*(ij2)} = \ell n \frac{x(ij.)x(i.1)x(i..)}{x(i..)x(ij.)x(i.2)} = \ell n \frac{x(i.1)}{x(i.2)} \quad \text{(wheeze)}.$$

The values of $\ell n(x(i1.)/x(i2.))$ and $\ell n(x(i.1)/x(i.2))$ are given in Grizzle (1971, 1060) and the values of $x_d^*(ijk)$ are given in Table 2.

From Fig. 1 we have the parametric representation

$$\ell n \frac{x_d^*(i1k)}{x_d^*(i2k)} = \tau_1^B + \tau_{i1}^{AB} \; ; \; \ell n \frac{x_d^*(ij1)}{x_d^*(ij2)} = \tau_1^W + \tau_{i1}^{AW} , \; i = 1,2, \ldots ,8$$

$$\ell n \frac{x_d^*(91k)}{x_d^*(92k)} = \tau_1^B \; ; \; \ell n \frac{x_d^*(9j1)}{x_d^*(9j2)} = \tau_1^W .$$

The values of the natural parameters in the parametric representation of the logits are

$$\tau_1^B = -0.3196, \; \tau_1^W = -0.2263, \text{ and}$$

	τ_{i1}^{AB}	τ_{i1}^{AW}
1	-4.4762	-2.6512
2	-3.6872	-2.3380
3	-3.0106	-1.8714
4	-2.4191	-1.6241
i = 5	-1.8993	-1.1955
6	-1.4214	-0.8840
7	-0.7828	-0.5713
8	-0.4394	-0.3466
9	0	0

FIGURE 1

Loglinear Representation

	1	2	3	...	9	10	11	12	13	...	19	20	21	...	27	28	29
ijk	L	τ^A_1	τ^A_2	...	τ^A_8	τ^B_1	τ^W_1	τ^{AB}_{11}	τ^{AB}_{21}	...	τ^{AB}_{81}	τ^{AW}_{11}	τ^{AW}_{21}	...	τ^{AW}_{81}	τ^{BW}_{11}	τ^{ABW}_{111}
111	1	1				1	1	1				1				1	1
112	1	1				1		1									
121	1	1					1					1					
122	1	1															
211	1		1			1	1		1				1			1	
212	1		1			1			1								
221	1		1				1						1				
222	1		1														
311	1					1	1									1	
312	1					1											
321	1						1										
322	1																
411	1					1	1									1	
412	1					1											
421	1						1										
422	1																
511	1					1	1									1	
512	1					1											
521	1						1										
522	1																
611	1					1	1									1	
612	1					1											
621	1						1										
622	1																
711	1					1	1									1	
712	1					1											
721	1						1										
722	1																
811	1				1	1	1				1				1	1	
812	1				1	1					1						
821	1				1		1								1		
822	1				1												
911	1					1	1									1	
912	1					1											
921	1						1										
922	1																

30	...	36	37	38	39
τ_{211}^{ABW}	...	τ_{811}^{ABW}	τ^{AB}	τ^{AW}	τ^{ABW}
			1	1	1
			1		
				1	
1			7/8	7/8	1
			7/8		
				7/8	
			6/8	6/8	1
			6/8		
				6/8	
			5/8	5/8	1
			5/8		
				5/8	
			4/8	4/8	1
			4/8		
				4/8	
			3/8	3/8	1
			3/8		
				3/8	
			2/8	2/8	1
			2/8		
				2/8	
		1	1/8	1/8	
			1/8		
				1/8	

TABLE 1

Number of subjects responding for the two symptoms in terms of age group

x(ijk)

Breathlessness			Yes, j = 1		No, j = 2		
			Yes	No	Yes	No	
Wheeze			k = 1	k = 2	k = 1	k = 2	Total
	1	20 - 24	9	7	95	1841	1952
	2	25 - 29	23	9	105	1654	1791
Age	3	30 - 34	54	19	177	1863	2113
groups	4	35 - 39	121	48	257	2357	2783
(years)	i = 5	40 - 44	169	54	273	1778	2274
	6	45 - 49	269	88	324	1712	2393
	7	50 - 54	404	117	245	1324	2090
	8	55 - 59	406	152	225	967	1750
	9	60 - 64	372	106	132	526	1136

Data from Ashford and Sowden (1970).

TABLE 2

$x_d^*(ijk)$

	j = 1		j = 2	
	k = 1	k = 2	k = 1	k = 2
1	0.852	15.148	103.147	1832.851
2	2.287	29.713	125.713	1633.287
3	7.981	65.019	223.019	1816.980
4	22.954	146.046	355.045	2258.954
i = 5	43.345	179.655	398.655	1652.344
6	88.467	268.533	504.533	1531.466
7	161.784	359.215	487.216	1081.784
8	201.199	356.801	429.801	762.198
9	212.070	265.929	291.929	366.070

In particular, Grizzle's objective was to calculate two lines relating the marginal logits to age, that is, to estimate and test the hypothesis

$$\ell n \frac{x_d^*(i1k)}{x_d^*(i2k)} = \alpha_1 + i\beta_1; \; \ell n \frac{x_d^*(ij1)}{x_d^*(ij2)} = \alpha_2 + i\beta_2, \; i = 1, \dots, 9.$$

But this hypothesis implies that the first-order differences in logits across age groups is constant, or in view of the parametric representation, that the first-order differences in the natural parameters are constant. These chains of equalities permit us to express the parameters τ_{i1}^{AB}, τ_{i1}^{AW} in terms of τ_{11}^{AB} and τ_{11}^{AW} as

$$\tau_{i1}^{AB} = \frac{9 - i}{8} \tau_{11}^{AB}, \; \tau_{i1}^{AW} = \frac{9 - i}{8} \tau_{11}^{AW}, \; i = 1, \dots, 8.$$

These relations among the natural parameters mean that in the loglinear representation the terms

$$\dots \tau_{11}^{AB} T_{11}^{AB}(ijk) + \tau_{21}^{AB} T_{21}^{AB}(ijk) + \dots + \tau_{81}^{AB} T_{81}^{AB}(ijk) \dots$$

reduce to

$$\tau_{11}^{AB}(T_{11}^{AB}(ijk) + \tfrac{7}{8}T_{21}^{AB}(ijk) + \tfrac{6}{8}T_{31}^{AB}(ijk) + \dots + \tfrac{1}{8}T_{81}^{AB}(ijk))$$

and the terms

$$\dots \tau_{11}^{AW} T_{11}^{AW}(ijk) + \tau_{21}^{AW} T_{21}^{AW}(ijk) + \dots + \tau_{81}^{AW} T_{81}^{AW}(ijk) \dots$$

reduce to

$$\tau_{11}^{AW}(T_{11}^{AW}(ijk) + \tfrac{7}{8}T_{21}^{AW}(ijk) + \tfrac{6}{8}T_{31}^{AW}(ijk) + \dots + \tfrac{1}{8}T_{81}^{AW}(ijk)).$$

If we denote the MDI estimate satisfying logit linearity by x_m^* then its design matrix or loglinear representation is given by Columns 1-11, 37, 38 of Fig. 1, where we use τ^{AB} and τ^{AW} respectively instead of τ_{11}^{AB} and τ_{11}^{AW}. The values of the MDI estimate x_m^* were determined by an iterative procedure subject to the constraints

$$x^*_m(i..) = x(i..),\ x^*_m(.j.) = x(.j.),\ x^*_m(..k) = x(..k),$$

$$\sum_{i=1}^{8} ((9 - i)/8)x^*_m(i1.) = \sum_{i=1}^{8} ((9 - i)/8)x(i1.),$$

$$\sum_{i=1}^{8} ((9 - i)/8)x^*_m(i.1) = \sum_{i=1}^{8} ((9 - i)/8)x(i.1).$$

The values of $x^*_m(ijk)$ are given in Table 3. The values of the natural parameters appearing in the linear model of the logits are

$$\tau_1^B = -0.2098,\ \tau^{AB} = -4.0996,\ \tau_1^W = -0.1841,\ \tau^{AW} = -2.6068.$$

The corresponding values of the logit representation in terms of the α's and β's as used by Grizzle (1971) are obtained from

$$\begin{matrix} \alpha_1 + 9\beta_1 = \tau_1^B \\ \alpha_1 + \beta_1 = \tau_1^B + \tau^{AB} \end{matrix},\qquad \begin{matrix} \alpha_2 + 9\beta_2 = \tau_1^W \\ \alpha_2 + \beta_2 = \tau_1^W + \tau^{AW} \end{matrix},$$

or

$$\alpha_1 = -4.8219,\ \beta_1 = 0.5125,\ \alpha_2 = -3.1167,\ \beta_2 = 0.3259.$$

TABLE 3

$x^*_m(ijk)$

	j = 1		j = 2	
	k = 1	k = 2	k = 1	k = 2
1	1.497	24.391	111.365	1814.747
2	3.079	36.225	137.235	1614.459
3	8.037	68.253	214.554	1822.152
4	22.967	140.816	367.283	2251.931
i = 5	39.612	175.330	379.461	1679.595
6	84.650	270.466	485.742	1552.140
7	142.437	328.542	489.605	1129.413
8	214.641	357.415	441.955	735.987
9	230.975	277.656	284.884	342.486

We also note that

$$\text{Var}(\alpha_1) = \text{Var}(\tau_1^B) + (81/64)\ \text{Var}(\tau^{AB}) + (18/8)\ \text{Cov}(\tau_1^B, \tau^{AB})$$

$$\text{Var}(\beta_1) = (1/64)\ \text{Var}(\tau^{AB})$$

$$\text{Var}(\alpha_2) = \text{Var}(\tau_1^W) + (81/64)\ \text{Var}(\tau^{AW}) + (18/8)\ \text{Cov}(\tau_1^W, \tau^{AW})$$

$$\text{Var}(\beta_2) = (1/64)\ \text{Var}(\tau^{AW}).$$

The covariance matrix of the natural parameters (taus) for x_m^* is obtained as follows (a weighted version of the procedure in Kullback 1959, 217). Compute $\underset{\sim}{S} = \underset{\sim}{T}'\underset{\sim}{D}\underset{\sim}{T}$ where $\underset{\sim}{T}$ is the design matrix for the loglinear representation of x_m^*(columns 1-11, 37, 38 of Fig. 1), and $\underset{\sim}{D}$ is a diagonal matrix whose entries are the values of $x_m^*(ijk)$ in lexicographic order. Partition the matrix $\underset{\sim}{S}$ as

$$\begin{bmatrix} \underset{\sim}{S}_{11} & \underset{\sim}{S}_{12} \\ \underset{\sim}{S}_{21} & \underset{\sim}{S}_{22} \end{bmatrix} \text{where } S_{11} \text{ is } 1 \times 1,$$

then the covariance matrix of the taus is $(\underset{\sim}{S}_{22} - \underset{\sim}{S}_{21}\underset{\sim}{S}_{11}^{-1}\underset{\sim}{S}_{12})^{-1}$. For comparison we list the values as given by Grizzle (1971) and as computed from x_m^*.

	Grizzle (1971)	x_m^*
α_1:	-4.8174 ± 0.0848	-4.8219 ± 0.0835
β_1:	0.5123 ± 0.0124	0.5125 ± 0.0129
α_2:	-3.1135 ± 0.0558	-3.1167 ± 0.0549
β_2:	0.3253 ± 0.0090	0.3258 ± 0.0089

The associated analysis of information Table 4 provides a basis for tests of significance and goodness-of-fit.

TABLE 4

Analysis of Information

Component due to	Information	D.F.
Interaction (linear logit model)	$2I(x:x_m^*) = 3077.154$	23
Effect	$2I(x_d^*:x_m^*) = 25.300$	14
Interaction (marginal logits)	$2I(x:x_d^*) = 3051.854$	9

We infer from the values of $2I(x:x_m^*)$ and $2I(x:x_d^*)$ that neither x_m^* or x_d^* is a good estimate for the joint response data, that is, $2I(x:x_m^*)$ $(2I(x:x_d^*))$ is a measure of the goodness-of-fit of the linear logit model (marginal logit model) to the joint response data. $2I(x_d^*:x_m^*)$ is a measure of the effect of the relationship among the natural parameters $\tau_{11}^{AB}, \tau_{21}^{AB}, \ldots, \tau_{81}^{AB}$ and $\tau_{11}^{AW}, \tau_{21}^{AW}, \ldots, \tau_{81}^{AW}$ of $x_d^*(ijk)$ implied by the hypothesis of logit linearity. We remark that x_m^* and x_d^* correspond respectively to model 3 and 8 of Mantel and Brown (1973). We shall return to the question of finding a model providing an acceptable fit to the joint response data of Table 1 after considering data giving the prevalence of persistent cough and persistent phlegm amongst the same group of miners.

In Table 5 is given a 9 x 2 x 2 cross-classification of the same miners as in Table 1, but showing the combined prevalence of persistent cough and persistent phlegm. We denote the observed frequency in any cell by x(ijk) with

Variable		Index	1	2	3	4	...	9
Age Group	A	i	20-24	25-29	30-34	35-39	...	60-64
Cough	C	j	yes	no				
Phlegm	P	k	yes	no				

Since Table 5 has the same dimensions as Table 1 the design matrix and loglinear presentation in Fig. 1 and the loglinear representa-

TABLE 5

Combined prevalence of persistent cough and persistent phlegm in British coal miners in terms of age - all smokers without pneumoconiosis

x(ijk)

Cough			Yes, j = 1		No, j = 2		
			Yes	No	Yes	No	
Phlegm			k = 1	k = 2	k = 1	k = 2	Total
	1	20 - 24	77	29	66	1780	1952
	2	25 - 29	89	40	64	1598	1791
Age	3	30 - 34	145	75	80	1813	2113
groups	4	35 - 39	237	101	107	2338	2783
(years)	5	40 - 44	282	116	82	1794	2274
	i = 6	45 - 49	373	152	99	1769	2393
	7	50 - 54	430	158	95	1407	2090
	8	55 - 59	445	122	88	1095	1750
	9	60 - 64	321	87	61	667	1136
			2399	880	742	14261	18282

Data from Ashford, Morgan, Rae, Sowden (1970).

tion (1) for the x(ijk) values of Table 1 will be the same for the x(ijk) of Table 5 with the replacement of the superscripts B, W by C, P respectively.

To determine the significance of effects and whether or not there is second-order interaction we fit a sequence of nested models based on the marginals

$$H_a\text{: } x(i..),\ x(.jk),$$

$$H_b\text{: } x(.jk),\ x(ij.),$$

$$H_c\text{: } x(.jk),\ x(ij.),\ x(i.k),$$

and denote the corresponding MDI estimates by x^*_a, x^*_b, x^*_c respectively. We note that x^*_a and x^*_b have the explicit form $x^*_a(ijk) = x(i..)x(.jk)/n$, $x^*_b(ijk) = x(ij.)x(.jk)/x(.j.)$ but x^*_c cannot be explicitly represented as a product of marginals. H_a

is the null hypothesis that the incidence of cough and phlegm is homogeneous over the age groups. H_b is the null hypothesis that the incidence of phlegm is homogeneous over the age groups given the incidence of cough. H_c is the null hypothesis of no second-order interaction. The columns of Fig. 1 implied for the design matrix or loglinear representation of the three models are

$$H_a: \; 1\text{-}11,\ 28, \quad H_b: \; 1\text{-}19,\ 28, \quad H_c: \; 1\text{-}28.$$

The hypotheses may also be stated as implying that the natural parameters corresponding to the columns of Fig. 1 not used in the design matrix or for the loglinear representation are zero. Analysis of information Table 6 summarizes the results.

TABLE 6

Analysis of Information

Component due to	Information	D.F.
a) x(i..), x(.jk)	$2I(x:x_a^*) = 1259.090$	24
b) x(.jk), x(ij.)	$2I(x_b^*:x_a^*) = 1180.385$	8
	$2I(x:x_b^*) = 78.705$	16
c) x(.jk), x(ij.), x(i.k)	$2I(x_c^*:x_b^*) = 72.009$	8
	$2I(x:x_c^*) = 6.696$	8

From Table 6 we infer that the 8 natural parameters corresponding to columns 29-36 of Fig. 1 may be taken as zero. From Fig. 1 we see that the parametric representations of the log-odds or logits under the model of no second-order interaction are

$$\ell n \frac{x_c^*(i11)}{x_c^*(i21)} = \tau_1^C + \tau_{i1}^{AC} + \tau_{11}^{CP}, \quad \ell n \frac{x_c^*(i12)}{x_c^*(i22)} = \tau_1^C + \tau_{i1}^{AC},$$

$$\ell n \frac{x_c^*(i11)}{x_c^*(i12)} = \tau_1^P + \tau_{i1}^{AP} + \tau_{11}^{CP}, \quad \ell n \frac{x_c^*(i21)}{x_c^*(i22)} = \tau_1^P + \tau_{i1}^{AP}, \quad i = 1,2, \ldots ,9.$$

The values of x_c^* are given in Table 8.

The values of the natural parameters in the representation of the logits are

$$\tau_1^C = -2.0987,\ \tau_1^P = -2.4756,\ \tau_{11}^{CP} = 3.8500, \text{ and}$$

	τ_{i1}^{AC}	τ_{i1}^{AP}
1	-1.7955	-0.7132
2	-1.5083	-0.6904
3	-1.1155	-0.6729
4	-1.0052	-0.5734
i = 5	-0.5939	-0.5473
6	-0.3801	-0.4448
7	-0.1422	-0.3070
8	-0.1103	-0.0639
9	0	0

The covariance matrix of these 19 parameters has been computed, but is not given herein. We mention however that the variance of τ_{11}^{CP} is 0.003116 so that $X^2 = (3.85)^2/0.003116 = 4756.90$ is approximately a chi-square with one degree of freedom. We see in Analysis of Information Table 7 a verification of the fact that the association parameter τ_{11}^{CP} is very significantly different from zero.

TABLE 7

Analysis of Information

Component due to	Information	D.F.
e) x(ij.), x(i.k)	$2I(x:x_e^*) = 6273.746$	9
c) x(ij.), x(i.k), x(.jk)	$2I(x_c^*:x_e^*) = 6267.050$	1
	$2I(x:x_c^*) = 6.696$	8

We remark that H_e: x(ij.), x(i.k) represents the model that cough

TABLE 8

x_c^*

	j = 1		j = 2	
	k = 1	k = 2	k = 1	k = 2
1	69.919	36.096	73.078	1772.902
2	85.750	43.272	67.248	1594.727
3	147.105	72.942	77.893	1815.057
4	233.341	104.742	110.657	2334.258
i = 5	276.957	121.121	87.043	1788.881
6	376.455	148.600	95.546	1772.402
7	437.482	150.460	87.521	1414.543
8	446.480	120.414	86.522	1096.588
9	325.511	82.354	56.491	671.648

and phlegm are not associated given the age grouping. The corresponding estimate may be explicitly represented as $x_e^*(ijk) = x(ij.)x(i.k)/x(i..)$. $2I(x_c^*:x_e^*)$ tests the null hypothesis that $\tau_{11}^{CP} = 0$ and the value of $2I(x:x_c^*) = 6.696$, 8 D.F. implies that the association between cough and phlegm has the same value over all the age groupings.

We now examine the hypothesis that the logits of x_c^* vary linearly with age, that is, that successive differences of the logits across age are constant. As before we can express the natural parameters τ_{i1}^{AC}, τ_{i1}^{AP}, under this hypothesis in terms of τ_{11}^{AC} and τ_{11}^{AP} as

$$H_n: \quad \tau_{i1}^{AC} = ((9 - i)/8)\ \tau_{11}^{AC}, \quad \tau_{i1}^{AP} = ((9 - i)/8)\ \tau_{11}^{AP}, \quad i = 1, \ldots, 8.$$

Denote the MDI estimate satisfying logit linearity within the model of no second-order interaction by x_n^*, then the design matrix or loglinear representation corresponding to H_n is given by columns 1-11, 28, 37, 38 of Fig. 1, of course, with the replacement of the superscripts B, W by C, P respectively and the use of τ^{AC}, τ^{AP} instead of τ_{11}^{AC}, τ_{11}^{AP} respectively for convenience.

The values of x_n^* are given in Table 9. The values of the natural parameters in the logit representation under the logit

TABLE 9

x_n^*

	j = 1		j = 2	
	k = 1	k = 2	k = 1	k = 2
1	72.602	42.728	64.450	1772.220
2	90.057	48.170	63.589	1589.185
3	142.725	69.383	80.161	1820.731
4	250.478	110.668	111.899	2309.956
i = 5	269.945	108.401	95.926	1799.728
6	370.019	135.044	104.586	1783.352
7	414.629	137.531	93.218	1444.623
8	437.604	131.922	78.255	1102.219
9	350.939	96.152	49.918	638.992

linearity model,

$$\ell n\ (x_n^*(i11)/x_n^*(i21)) = \tau_1^C + ((9 - i)/8)\tau^{AC} + \tau_{11}^{CP},$$

$$\ell n\ (x_n^*(i12)/x_n^*(i22)) = \tau_1^C + ((9 - i)/8)\tau^{AC},$$

$$\ell n\ (x_n^*(i11)/x_n^*(i12)) = \tau_1^P + ((9 - i)/8)\tau^{AP} + \tau_{11}^{CP},$$

$$\ell n\ (x_n^*(i21)/x_n^*(i22)) = \tau_1^P + ((9 - i)/8)\tau^{AP},$$

are

$$\tau_1^C = -1.8939,\ \tau_1^P = -2.5495,$$

$$\tau^{AC} = -1.8312,\ \tau^{AP} = -0.7646,\ \tau_{11}^{CP} = 3.8442.$$

The covariance matrix of these five parameters is given in Table 10. The associated analysis of information is given in Table 11.

TABLE 10

Covariance matrix of τ_1^C, τ_1^P, τ^{AC}, τ^{AP}, τ_{11}^{CP} values in x_n^*

τ_1^C	τ_1^P	τ^{AC}	τ^{AP}	τ_{11}^{CP}
0.0028	-0.0011	-0.0038	0.0024	-0.0011
	0.0037	0.0029	-0.0046	-0.0019
		0.0091	-0.0060	-0.0004
			0.0092	0.0010
				0.0031

TABLE 11

Analysis of Information

Component due to	Information	D.F.
H_n	$2I(x:x_n^*) = 28.831$	22
H_c	$2I(x_c^*:x_n^*) = 22.135$	14
	$2I(x:x_c^*) = 6.696$	8

The value $2I(x:x_n^*)$ is a measure of the goodness-of-fit of the logit linearity model and $2I(x_c^*:x_n^*)$ is a measure of the effect of replacing the common natural parameters τ^{AC}, τ^{AP} by τ_{i1}^{AC}, τ_{i1}^{AP}, $i = 1, \ldots, 8$. It is clear that x_c^* provides a better fit to the original data than x_n^*, using more parameters however, but at the 5% level of significance the logit linearity model provides an acceptable fit, with a simpler model.

In our analysis of the incidence of cough and phlegm over the age groups we concluded that the association of these factors was the same over all the age groupings. However, in multidimensional contingency tables in which, for example, time or age is one of the classifications, there may occur an age effect such that an hypothesis of interest may be rejected for the entire table, but an hypothesis taking the possible age effect into account may produce an acceptable partitioning. We now propose to illustrate techniques applicable to the solution of such problems by a further study of the 9 x 2 x 2 contingency Table 1, containing nine age groupings, for which the hypothesis of no second-order interaction is rejected. An acceptable partitioning is determined. Within the partitioned model we then consider a subhypothesis of logit linearity.

TABLE 12

No second-order interaction estimate for the data of Table 1

$x_2^*(ijk)$

	j = 1		j = 2	
	k = 1	k = 2	k = 1	k = 2
1	7.547	8.454	96.448	1839.547
2	17.089	14.914	110.907	1648.087
3	45.954	27.054	185.040	1854.947
4	111.407	57.611	266.585	2347.390
i = 5	162.527	60.504	279.467	1771.497
6	271.823	85.231	321.175	1714.769
7	398.159	122.871	250.848	1318.129
8	431.692	126.271	199.319	992.729
9	380.802	97.091	123.210	534.909

$$\ln \frac{x_2^*(i11)x_2^*(i22)}{x_2^*(i12)x_2^*(i21)} = \tau_{11}^{BW} = 2.8348$$

Let us now find the MDI estimate for the data in Table 1 under the classic null hypothesis of no second-order interaction. The MDI estimate $x_2^*(ijk)$ under the hypothesis H_2 of no second-order interaction is obtained by iteratively fitting the marginals x(ij.), x(i.k), x(.jk) and is given in Table 12. The design matrix or loglinear representation of $x_2^*(ijk)$ is given by the columns 1-28 in Fig. 1. Indeed, the no second-order interaction hypothesis is that the values of the last eight natural parameters in the complete loglinear representation of x(ijk) have the hypothetical values

(2) $$\tau_{111}^{ABW} = \tau_{211}^{ABW} = \dots = \tau_{811}^{ABW} = 0.$$

Computing the associated MDI statistic we find

$$2I(x:x_2^*) = 2\Sigma\Sigma\Sigma x(ijk)\ \ell n(x(ijk)/x_2^*(ijk)) = 26.673,\ 8\ \text{D.F.}$$

We recall that this is the same as the log-likelihood ratio chi-square statistic (see e.g. Darroch 1962). We reject the null hypothesis of no second-order interaction, that is, the hypothetical values in (2) are not acceptable natural parameter values for x(ijk).

Among other properties, the null hypothesis of no second-order interaction implies a common value for the association (measured by the logarithm of the cross-product ratio) between breathlessness and wheeze over all age-groups. In terms of the natural parameters defining $x_2^*(ijk)$ this common value as determined from columns 1-28 of Fig. 1 is

$$\ell n \frac{x_2^*(i11)x_2^*(i22)}{x_2^*(i12)x_2^*(i21)} = \tau_{11}^{BW} = 2.8348,\ i = 1,2,\ \dots\ ,9.$$

We summarize the results and supplement analysis of information Table 4 by analysis of information Table 13.

TABLE 13

Analysis of Information

Component due to	Information	D.F.
d) x(ij.), x(i.k)	$2I(x:x_d^*) = 3051.854$	9
H_2: x(ij.), x(i.k), x(.jk)	$2I(x_2^*:x_d^*) = 3025.181$	1
	$2I(x:x_2^*) = 26.673$	8

The value of $2I(x_2^*:x_d^*)$ implies a significant (nonzero) association between breathlessness and wheeze but the value of $2I(x:x_2^*)$ leads one to conclude that there is not a common value of this association over all the age groups. We note that the MDI estimate x_2^* corresponds to model 9 of Mantel and Brown (1973).

It seems reasonable to conjecture that the presence of second-order interaction may be related to an age effect. That is, there may be a common value of the association between breathlessness and wheeze over some of the younger age groups and a common but different value of this association over the remaining age groups. We therefore re-examined the computer output for x_2^*. Among other items there was given for each cell a number called OUTLIER, the value of

$$2(x(ijk)\ \ell n(x(ijk)/x_2^*(ijk)) + (n - x(ijk))\ell n\ (n - x(ijk))/(n - x_2^*(ijk))).$$

Ireland (1972) has shown that large values of OUTLIER are effective in recognizing outliers under the estimation procedure in question. In the case at hand the value of OUTLIER for cell 812 was 4.959 with the next largest value 2.722 for cell 212.

Let us therefore consider a partitioning of the second-order interaction for the age groups under 55 and for the age groups 55 and over by computing the MDI estimate $x_t^*(ijk)$ under the marginal constraints of $x_2^*(ijk)$ and also the constraints

$$(3) \qquad \tau_{111}^{ABW} = \tau_{211}^{ABW} = \dots = \tau_{711}^{ABW},\ \tau_{811}^{ABW} = \tau_{911}^{ABW} = 0.$$

The design matrix or loglinear representation for $x_t^*(ijk)$ is given by columns 1-28, 39 in Fig. 1, that is, with the eight columns corresponding to $\tau_{111}^{ABW}, \tau_{211}^{ABW}, \dots, \tau_{811}^{ABW}$ replaced by the one column labeled τ^{ABW}. The values of $x_t^*(ijk)$ are given in Table 14. In terms of the natural parameters defining $x_t^*(ijk)$, from columns 1-28, 39 in Fig. 1, it is found that

$$\ell n\ \frac{x_t^*(i11)x_t^*(i22)}{x_t^*(i12)x_t^*(i21)} = \tau_{11}^{BW} + \tau^{ABW} = 3.0007,\ i = 1, \dots ,7$$

$$\ell n\ \frac{x_t^*(i11)x_t^*(i22)}{x_t^*(i12)x_t^*(i21)} = \tau_{11}^{BW} \qquad = 2.5212,\ i = 8,9.$$

TABLE 14

Partitioned second-order interaction estimate

$x_t^*(ijk)$

	j = 1		j = 2	
	k = 1	k = 2	k = 1	k = 2
i = 1	8.182	7.819	95.816	1840.183
2	18.306	13.695	109.692	1649.306
3	48.466	24.539	182.532	1857.463
4	116.719	52.292	261.279	2352.709
5	168.521	54.497	273.479	1777.504
6	280.217	76.810	312.784	1723.192
7	408.590	112.349	240.417	1328.652
8	411.545	146.550	219.454	972.450
9	366.455	111.450	137.546	520.550

$$\ell n \frac{x_t^*(i11)x_t^*(i22)}{x_t^*(i12)x_t^*(i21)} = 3.0007, \ i = 1, \ldots, 7$$

$$\ell n \frac{x_t^*(i11)x_t^*(i22)}{x_t^*(i12)x_t^*(i21)} = 2.5212, \ i = 8,9$$

The associated analysis of information Table 15 summarizes results.

TABLE 15

Analysis of Information

Component due to	Information	D.F.
No second-order interaction	$2I(x:x_2^*) = 26.673$	8
Effect	$2I(x_t^*:x_2^*) = 16.700$	1
Interaction (partition)	$2I(x:x_t^*) = 9.973$	7

We note that $2I(x_t^*:x_2^*)$ which measures the effect of the hypothesis in (3) is very significant, and from the value of $2I(x:x_t^*)$ we may accept the inference that there is a common association between breathlessness and wheeze for the age groups under 55 and a different but common value for the age groups 55 and over and that in fact $x_t^*(ijk)$ is a good fit to the original data. We remark that, as a matter of fact, the values of $x_t^*(ijk)$ were computed by iteratively fitting all the two-way marginals of the 7 x 2 x 2 table of the age groups under 55 and separately iteratively fitting all the two-way marginals of the 2 x 2 x 2 table of the age groups 55 and over.

To verify the indication given by OUTLIER we also examined the other possible "break points" with the following results

Partition	$2I(x:x_2^*)$	D.F.
Under 35	0.612	2
Over 35	15.990	5
Under 40	1.856	3
Over 40	11.541	4
Under 45	3.311	4
Over 45	8.373	3
Under 50	8.420	5
Over 50	7.861	2

These values confirm the inference suggested by OUTLIER.

If we now consider the logits for breathlessness and wheeze, respectively, for the age groups under 55, from the design matrix or loglinear representation for $x_t^*(ijk)$ in Fig. 1 (columns 1-28, 39) we see that

$$\ell n\ x_t^*(i11)/x_t^*(i21) = \tau_1^B + \tau_{i1}^{AB} + \tau_{11}^{BW} + \tau^{ABW};$$

$$\ell n\ x_t^*(i12)/x_t^*(i22) = \tau_1^B + \tau_{i1}^{AB},\ i = 1, \ldots, 7$$

$$\ell n\ x_t^*(i11)/x_t^*(i12) = \tau_1^W + \tau_{i1}^{AW} + \tau_{11}^{BW} + \tau^{ABW};$$

$$\ell n\ x_t^*(i21)/x_t^*(i22) = \tau_1^W + \tau_{i1}^{AW},\ i = 1, \ldots, 7.$$

The corresponding logits for the age groups 55 and over are given by

$$\ln \frac{x_t^*(811)}{x_t^*(821)} = \tau_1^B + \tau_{81}^{AB} + \tau_{11}^{BW}; \quad \ln \frac{x_t^*(812)}{x_t^*(822)} = \tau_1^B + \tau_{81}^{AB}$$

$$\ln \frac{x_t^*(911)}{x_t^*(921)} = \tau_1^B + \tau_{11}^{BW}; \quad \ln \frac{x_t^*(912)}{x_t^*(922)} = \tau_1^B$$

$$\ln \frac{x_t^*(811)}{x_t^*(812)} = \tau_1^W + \tau_{81}^{AW} + \tau_{11}^{BW}; \quad \ln \frac{x_t^*(821)}{x_t^*(822)} = \tau_1^W + \tau_{81}^{AW}$$

$$\ln \frac{x_t^*(911)}{x_t^*(912)} = \tau_1^W + \tau_{11}^{BW}; \quad \ln \frac{x_t^*(921)}{x_t^*(922)} = \tau_1^W.$$

The numerical values of these logits are given in Table 16.

TABLE 16

Logits

	Breathlessness $\ln \frac{x_t^*(i1k)}{x_t^*(i2k)}$ k = 1	k = 2	Wheeze $\ln \frac{x_t^*(ij1)}{x_t^*(ij2)}$ j = 1	j = 2
i = 1	-2.4605	-5.4611	0.0455	-2.9552
2	-1.7904	-4.7911	0.2902	-2.7104
3	-1.3261	-4.3267	0.6806	-2.3200
4	-0.8058	-3.8065	0.8029	-2.1977
5	-0.4842	-3.4848	1.1289	-1.8717
6	-0.1100	-3.1106	1.2942	-1.7064
7	0.5303	-2.4703	1.2911	-1.7095
8	0.6288	-1.8925	1.0326	-1.4887
9	0.9799	-1.5413	1.1903	-1.3309

We now consider the hypothesis that within the partitioned no second-order hypothesis, that is, within the $x_t^*(ijk)$ model, the logits are linearly related for the age groups under 55, in other words, we consider the fitting of straight lines to the logits for the age groups under 55 by assuming that the differences of logits for successive age groups are constant. Thus we shall consider a null hypothesis that

$$\tau_{71}^{AB} - \tau_{61}^{AB} = \tau_{61}^{AB} - \tau_{51}^{AB} = \tau_{51}^{AB} - \tau_{41}^{AB} = \dots = \tau_{21}^{AB} - \tau_{11}^{AB},$$

$$\tau_{71}^{AW} - \tau_{61}^{AW} = \tau_{61}^{AW} - \tau_{51}^{AW} = \tau_{51}^{AW} - \tau_{41}^{AW} = \dots = \tau_{21}^{AW} - \tau_{11}^{AW}.$$

If, as a matter of convenience, we consider the design matrix or loglinear representation of $x_t^*(ijk)$ as in Fig. 2, that is, a reparametrization of the loglinear representation in Fig. 1, then the chains of equalities yield the relations among the natural parameters

$$\tau_{i1}^{AB} = ((7 - i)/6)\tau_{11}^{AB},\ \tau_{i1}^{AW} = ((7 - i)/6)\tau_{11}^{AW},\ i = 1,2, \dots ,7.$$

The design matrix or loglinear representation for the linear logit model MDI estimate $x_v^*(ijk)$, using τ^{AB} and τ^{AW} respectively, instead of τ_{11}^{AB} and τ_{11}^{AW} is given in columns 1-11, 28-31 of Fig. 2. The values in columns 30, 31 arise from the fact that in the loglinear representation as in (1) the terms

$$\tau_{11}^{AB}T_{11}^{AB}(ijk) + \tau_{21}^{AB}T_{21}^{AB}(ijk) + \dots + \tau_{61}^{AB}T_{61}^{AB}(ijk)$$

and the terms

$$\tau_{11}^{AW}T_{11}^{AW}(ijk) + \tau_{21}^{AW}T_{21}^{AW}(ijk) + \dots + \tau_{61}^{AW}T_{61}^{AW}(ijk)$$

because of the relations among the natural parameters reduce to

$$\tau^{AB}(T_{11}^{AB}(ijk) + (5/6)T_{21}^{AB}(ijk) + (4/6)T_{31}^{AB}(ijk) + \dots + (1/6)T_{61}^{AB}(ijk))$$

and

$$\tau^{AW}(T_{11}^{AW}(ijk) + (5/6)T_{21}^{AW}(ijk) + (4/6)T_{31}^{AW}(ijk) + \dots + (1/6)T_{61}^{AW}(ijk))$$

respectively. The iteration used to compute $x_v^*(ijk)$ is (see Chapter 6 Section V)

FIGURE 2

Loglinear Representation

	1	2	3	4	5	6	7	8	9	10	11	12	13	14	15	16	17
ijk	L	τ_1^A	τ_2^A	τ_3^A	τ_4^A	τ_5^A	τ_6^A	τ_8^A	τ_9^A	τ_1^B	τ_1^W	τ_{11}^{AB}	τ_{21}^{AB}	τ_{31}^{AB}	τ_{41}^{AB}	τ_{51}^{AB}	τ_{61}^{AB}
111	1	1								1	1	1					
112	1	1								1		1					
121	1	1									1						
122	1	1															
211	1		1							1	1		1				
212	1		1							1			1				
221	1		1								1						
222	1		1														
311	1			1						1	1			1			
312	1			1						1				1			
321	1			1							1						
322	1			1													
411	1				1					1	1				1		
412	1				1					1					1		
421	1				1						1						
422	1				1												
511	1					1				1	1					1	
512	1					1				1						1	
521	1					1					1						
522	1					1											
611	1						1			1	1						1
612	1						1			1							1
621	1						1				1						
622	1						1										
711	1									1	1						
712	1									1							
721	1										1						
722	1																
811	1							1		1	1						
812	1							1		1							
821	1							1			1						
822	1							1									
911	1								1	1	1						
912	1								1	1							
921	1								1		1						
922	1								1								

18	19	20	21	22	23	24	25	26	27	28	29	30	31
τ_{81}^{AB}	τ_{91}^{AB}	τ_{11}^{AW}	τ_{21}^{AW}	τ_{31}^{AW}	τ_{41}^{AW}	τ_{51}^{AW}	τ_{61}^{AW}	τ_{81}^{AW}	τ_{91}^{AW}	τ_{11}^{BW}	τ^{ABW}	τ^{AB}	τ^{AW}
		1								1		1	1
												1	
		1											1
			1							1		5/6	5/6
												5/6	
			1										5/6
				1						1		4/6	4/6
												4/6	
				1									4/6
					1					1		3/6	3/6
												3/6	
					1								3/6
						1				1		2/6	2/6
												2/6	
						1							2/6
							1			1		1/6	1/6
												1/6	
							1						1/6
										1			
1								1		1	1		
1													
								1					
	1								1	1	1		
	1												
									1				

$$x^{(5n+1)}(ijk) = \frac{x(i..)}{x^{(5n)}(i..)} x^{(5n)}(ijk)$$

$$x^{(5n+2)}(ijk) = \frac{x(.j.)}{x^{(5n+1)}(.j.)} x^{(5n+1)}(ijk)$$

$$x^{(5n+3)}(ijk) = \frac{x(..k)}{x^{(5n+2)}(..k)} x^{(5n+2)}(ijk)$$

$$x^{(5n+4)}(ijk) = \left(\frac{h_1}{h_1^{(5n+3)}}\right)^{a_1(ijk)} \left(\frac{h_2}{h_2^{(5n+3)}}\right)^{a_2(ijk)} \left(\frac{h_3}{h_3^{(5n+3)}}\right)^{a_3(ijk)} x^{(5n+3)}(ijk)$$

$$x^{(5n+5)}(ijk) = \left(\frac{k_1}{k_1^{(5n+4)}}\right)^{b_1(ijk)} \left(\frac{k_2}{k_2^{(5n+4)}}\right)^{b_2(ijk)} \left(\frac{k_3}{k_3^{(5n+4)}}\right)^{b_3(ijk)} x^{(5n+4)}(ijk)$$

$$x^{(0)}(ijk) = n/28, \ n = \sum_{i=1}^{7} \sum_{j=1}^{2} \sum_{k=1}^{2} x(ijk).$$

All marginals refer to the 7 x 2 x 2 table and the values of $a_m(ijk)$, $b_m(ijk)$, $m = 1,2,3$ and the constraints $h_m, k_m, m = 1,2,3$ are given in Fig. 3. We remark that since $x_v^*(ijk) = x_t^*(ijk)$ for $i = 8,9$, we can perform the iteration by consideration of the 7 x 2 x 2 table only. The values of $x_v^*(ijk)$ are given in Table 17. For another example using this iteration see Kullback and Cornfield (1976).

Results are summarized in analysis of information Table 18.

TABLE 17

Linear logit estimate within partitioned second-order interaction model

$x_v^*(ijk)$

	j = 1		j = 2	
	k = 1	k = 2	k = 1	k = 2
i = 1	11.860	9.990	108.934	1821.215
2	20.398	13.952	120.522	1636.127
3	44.705	24.830	169.946	1873.519
4	107.932	48.677	263.913	2362.476
5	158.232	57.944	248.880	1808.943
6	288.909	85.919	292.375	1725.797
7	416.964	100.688	271.429	1300.919
8	411.545	146.550	219.454	972.450
9	366.455	111.450	137.546	520.550

$$\ln \frac{x_v^*(i11)x_v^*(i22)}{x_v^*(i12)x_v^*(i21)} = 2.9881, \quad i = 1, \ldots, 7$$

$$\ln \frac{x_v^*(i11)x_v^*(i22)}{x_v^*(i12)x_v^*(i21)} = 2.5212, \quad i = 8,9$$

TABLE 18

Analysis of Information

Component due to	Information	D.F.
Interaction (linear logits)	$2I(x:x_v^*) = 29.560$	17
Effect	$2I(x_t^*:x_v^*) = 19.587$	10
Interaction (partition)	$2I(x:x_t^*) = 9.973$	7

FIGURE 3

τ^{AB}

	111	112	121	122	211	212	221	222	311	312	321	322	411	412
$a_1(ijk)$	1	1			5/6	5/6			4/6	4/6			3/6	3/6
$a_2(ijk)$			1	1			5/6	5/6			4/6	4/6		
$a_3(ijk)$					1/6	1/6	1/6	1/6	2/6	2/6	2/6	2/6	3/6	3/6

τ^{AW}

	111	112	121	122	211	212	221	222	311	312	321	322	411	412
$b_1(ijk)$	1		1		5/6		5/6		4/6		4/6		3/6	
$b_2(ijk)$		1		1		5/6		5/6		4/6		4/6		3/6
$b_3(ijk)$					1/6	1/6	1/6	1/6	2/6	2/6	2/6	2/6	3/6	3/6

h_1: $x(11.) + (5/6)x(21.) + (4/6)x(31.) + (3/6)x(41.)$

h_2: $x(12.) + (5/6)x(22.) + (4/6)x(32.) + (3/6)x(42.)$

h_3: $(1/6)x(2..) + (2/6)x(3..) + (3/6)x(4..)$

k_1: $x(1.1) + (5/6)x(2.1) + (4/6)x(3.1) + (3/6)x(4.1)$

k_2: $x(1.2) + (5/6)x(2.2) + (4/6)x(3.2) + (3/6)x(4.2)$

k_3: $(1/6)x(2..) + (2/6)x(3..) + (3/6)x(4..)$

All marginals refer to the 7 x 2 x 2 table for age groups under 55.

τ^{AB}

421	422	511	512	521	522	611	612	621	622	711	712	721	722
		2/6	2/6			1/6	1/6						
3/6	3/6			2/6	2/6			1/6	1/6				
3/6	3/6	4/6	4/6	4/6	4/6	5/6	5/6	5/6	5/6	1	1	1	1

τ^{AW}

3/6		2/6		2/6		1/6		1/6					
	3/6		2/6		2/6		1/6		1/6				
3/6	3/6	4/6	4/6	4/6	4/6	5/6	5/6	5/6	5/6	1	1	1	1

+ (2/6)x(51.) + (1/6)x(61.)

+ (2/6)x(52.) + (1/6)x(62.)

+ (4/6)x(5..) + (5/6)x(6..) + x(7..)

+ (2/6)x(5.1) + (1/6)x(6.1)

+ (2/6)x(5.2) + (1/6)x(6.2)

+ (4/6)x(5..) + (5/6)x(6..) + x(7..)

Since $2I(x:x_v^*)$ and $2I(x_t^*:x_v^*)$ fall between the 5% and 2% values of the tabulated chi-square values with the appropriate degrees of freedom, we might accept the null hypothesis of linearity of the logits within the partitioned second-order interaction model, that is, infer from the value of $2I(x_t^*:x_v^*)$ that the natural parameters $\tau_{11}^{AB}, \tau_{21}^{AB}, \ldots, \tau_{71}^{AB}$ and $\tau_{11}^{AW}, \tau_{21}^{AW}, \ldots, \tau_{71}^{AW}$ of $x_t^*(ijk)$ satisfy the relations among the natural parameters implied by the logit linearity and that the MDI estimate $x_v^*(ijk)$ under the logit linearity model is an acceptable estimate for the original observations.

We have seen that a possible alternative to the hypothesis in (2), as suggested by OUTLIER, is the hypothesis in (3). Another possible alternative is suggested by the following observation (Plackett, 1974, 95). The associations between breathlessness and wheeze for the nine age groups in the observed data are (see Table 1):

1. $\ln(9 \times 1841/7 \times 95) = 3.216$
2. $\ln(23 \times 1654/9 \times 105) = 3.695$
3. $\ln(54 \times 1863/19 \times 177) = 3.398$
4. $\ln(121 \times 2357/48 \times 257) = 3.141$
5. $\ln(169 \times 1778/54 \times 273) = 3.015$
6. $\ln(269 \times 1712/88 \times 324) = 2.782$
7. $\ln(404 \times 1324/117 \times 245) = 2.926$
8. $\ln(406 \times 967/152 \times 225) = 2.441$
9. $\ln(372 \times 526/106 \times 132) = 2.638$

These suggest that the associations may vary linearly with age. Now for the observed data the log-odds or logits of breathlessness and wheeze are respectively given by

$$\ln(x(i11)/x(i21)) = \tau_1^B + \tau_{i1}^{AB} + \tau_{11}^{BW} + \tau_{i11}^{ABW},$$

$$\ln(x(i12)/x(i22)) = \tau_1^B + \tau_{i1}^{AB},$$

or

$$\ell n(x(i11)x(i22)/x(i21)x(i12)) = \tau_{11}^{BW} + \tau_{i11}^{ABW},$$

and

$$\ell n(x(i11)/x(i12)) = \tau_{1}^{W} + \tau_{i1}^{AW} + \tau_{11}^{BW} + \tau_{i11}^{ABW},$$

$$\ell n(x(i21)/x(i22)) = \tau_{1}^{W} + \tau_{i1}^{AW},$$

or

$$\ell n(x(i11)x(i22)/x(i12)x(i21)) = \tau_{11}^{BW} + \tau_{i11}^{ABW}.$$

Thus to obtain the MDI estimate for which the associations vary linearly with age, or the differences of the associations for successive age groups are constant, we want the MDI estimate satisfying the constraints for x_2^* (fitting all the two-way marginals) and in addition satisfying the constraint $\tau_{i11}^{ABW} = (9 - i)\tau^{ABW}$. In order to obtain the appropriate MDI estimate x_w^*, we shall use the k-sample Newton-Raphson type iteration described in Chapter 5, treating the data as nine sets of four cell multinomials. The appropriate B-matrix for x_w^* is given in Table 19. The values of the weights to derive the C-matrix and the values of the constraints $\underset{\sim}{C}\underset{\sim}{x}_w^* = \underset{\sim}{C}\underset{\sim}{x}$ are given in Table 20. The values of the MDI estimate x_w^* are given in Table 21. The analysis of information for the MDI estimate x_w^* is summarized in Table 22. Note that the B-matrix for the k-sample iteration for the no second-order interaction MDI estimate x_2^* consists of rows 1 to 28 of Table 19. It is seen that the MDI estimate x_w^* is a somewhat better fit than the interaction partition MDI estimate x_t^* (see Table 15), which used a step function rather than a linear function.

TABLE 19

B-matrix for x^*_w

A i	1	1	1	1	2	2	2	2	3	3	3	3	4	4	4	4	5	5	5	5	6	6	6	6	7	7	7	7
B j	1	1	2	2	1	1	2	2	1	1	2	2	1	1	2	2	1	1	2	2	1	1	2	2	1	1	2	2
W k	1	2	1	2	1	2	1	2	1	2	1	2	1	2	1	2	1	2	1	2	1	2	1	2	1	2	1	2
1	1	1	1	1	0	0	0	0	0	0	0	0	0	0	0	0	0	0	0	0	0	0	0	0	0	0	0	0
2	0	0	0	0	1	1	1	1	0	0	0	0	0	0	0	0	0	0	0	0	0	0	0	0	0	0	0	0
3	0	0	0	0	0	0	0	0	1	1	1	1	0	0	0	0	0	0	0	0	0	0	0	0	0	0	0	0
4	0	0	0	0	0	0	0	0	0	0	0	0	1	1	1	1	0	0	0	0	0	0	0	0	0	0	0	0
5	0	0	0	0	0	0	0	0	0	0	0	0	0	0	0	0	1	1	1	1	0	0	0	0	0	0	0	0
6	0	0	0	0	0	0	0	0	0	0	0	0	0	0	0	0	0	0	0	0	1	1	1	1	0	0	0	0
7	0	0	0	0	0	0	0	0	0	0	0	0	0	0	0	0	0	0	0	0	0	0	0	0	1	1	1	1
8	0	0	0	0	0	0	0	0	0	0	0	0	0	0	0	0	0	0	0	0	0	0	0	0	0	0	0	0
9	0	0	0	0	0	0	0	0	0	0	0	0	0	0	0	0	0	0	0	0	0	0	0	0	0	0	0	0
10	1	1	0	0	1	1	0	0	1	1	0	0	1	1	0	0	1	1	0	0	1	1	0	0	1	1	0	0
11	1	0	1	0	1	0	1	0	1	0	1	0	1	0	1	0	1	0	1	0	1	0	1	0	1	0	1	0
12	1	1	0	0	0	0	0	0	0	0	0	0	0	0	0	0	0	0	0	0	0	0	0	0	0	0	0	0
13	0	0	0	0	1	1	0	0	0	0	0	0	0	0	0	0	0	0	0	0	0	0	0	0	0	0	0	0
14	0	0	0	0	0	0	0	0	1	1	0	0	0	0	0	0	0	0	0	0	0	0	0	0	0	0	0	0
15	0	0	0	0	0	0	0	0	0	0	0	0	1	1	0	0	0	0	0	0	0	0	0	0	0	0	0	0
16	0	0	0	0	0	0	0	0	0	0	0	0	0	0	0	0	1	1	0	0	0	0	0	0	0	0	0	0
17	0	0	0	0	0	0	0	0	0	0	0	0	0	0	0	0	0	0	0	0	1	1	0	0	0	0	0	0
18	0	0	0	0	0	0	0	0	0	0	0	0	0	0	0	0	0	0	0	0	0	0	0	0	1	1	0	0
19	0	0	0	0	0	0	0	0	0	0	0	0	0	0	0	0	0	0	0	0	0	0	0	0	0	0	0	0
20	1	0	1	0	0	0	0	0	0	0	0	0	0	0	0	0	0	0	0	0	0	0	0	0	0	0	0	0
21	0	0	0	0	1	0	1	0	0	0	0	0	0	0	0	0	0	0	0	0	0	0	0	0	0	0	0	0
22	0	0	0	0	0	0	0	0	1	0	1	0	0	0	0	0	0	0	0	0	0	0	0	0	0	0	0	0
23	0	0	0	0	0	0	0	0	0	0	0	0	1	0	1	0	0	0	0	0	0	0	0	0	0	0	0	0
24	0	0	0	0	0	0	0	0	0	0	0	0	0	0	0	0	1	0	1	0	0	0	0	0	0	0	0	0
25	0	0	0	0	0	0	0	0	0	0	0	0	0	0	0	0	0	0	0	0	1	0	1	0	0	0	0	0
26	0	0	0	0	0	0	0	0	0	0	0	0	0	0	0	0	0	0	0	0	0	0	0	0	1	0	1	0
27	0	0	0	0	0	0	0	0	0	0	0	0	0	0	0	0	0	0	0	0	0	0	0	0	0	0	0	0
28	1	0	0	0	1	0	0	0	1	0	0	0	1	0	0	0	1	0	0	0	1	0	0	0	1	0	0	0
29	8	0	0	0	7	0	0	0	6	0	0	0	5	0	0	0	4	0	0	0	3	0	0	0	2	0	0	0

8	8	8	8	9	9	9	9	
1	1	2	2	1	1	2	2	
1	2	1	2	1	2	1	2	
0	0	0	0	0	0	0	0	
0	0	0	0	0	0	0	0	
0	0	0	0	0	0	0	0	
0	0	0	0	0	0	0	0	
0	0	0	0	0	0	0	0	
0	0	0	0	0	0	0	0	
0	0	0	0	0	0	0	0	
1	1	1	1	0	0	0	0	
0	0	0	0	1	1	1	1	
1	1	0	0	1	1	0	0	τ_1^B
1	0	1	0	1	0	1	0	τ_1^W
0	0	0	0	0	0	0	0	τ_{11}^{AB}
0	0	0	0	0	0	0	0	τ_{21}^{AB}
0	0	0	0	0	0	0	0	τ_{31}^{AB}
0	0	0	0	0	0	0	0	τ_{41}^{AB}
0	0	0	0	0	0	0	0	τ_{51}^{AB}
0	0	0	0	0	0	0	0	τ_{61}^{AB}
0	0	0	0	0	0	0	0	τ_{71}^{AB}
1	1	0	0	0	0	0	0	τ_{81}^{AB}
0	0	0	0	0	0	0	0	τ_{11}^{AW}
0	0	0	0	0	0	0	0	τ_{21}^{AW}
0	0	0	0	0	0	0	0	τ_{31}^{AW}
0	0	0	0	0	0	0	0	τ_{41}^{AW}
0	0	0	0	0	0	0	0	τ_{51}^{AW}
0	0	0	0	0	0	0	0	τ_{61}^{AW}
0	0	0	0	0	0	0	0	τ_{71}^{AW}
1	0	1	0	0	0	0	0	τ_{81}^{AW}
1	0	0	0	1	0	0	0	τ_{11}^{BW}
1	0	0	0	0	0	0	0	τ^{ABW}

TABLE 20

The weights to derive $\underset{\sim}{C}$ are

$w_1 = 0.106772$	$w_4 = 0.152226$	$w_7 = 0.114320$
$w_2 = 0.097965$	$w_5 = 0.124385$	$w_8 = 0.095723$
$w_3 = 0.115578$	$w_6 = 0.120894$	$w_9 = 0.062138$

The values of the constraints $\underset{\sim}{C}\underset{\sim}{x}^*_w = \underset{\sim}{C}\underset{\sim}{x}$ are

(1)	18282	(10)	2427	(20)	104
(2)	18282	(11)	3660	(21)	128
(3)	18282	(12)	16	(22)	231
(4)	18282	(13)	32	(23)	378
(5)	18282	(14)	73	(24)	442
(6)	18282	(15)	169	(25)	593
(7)	18282	(16)	223	(26)	649
(8)	18282	(17)	357	(27)	631
(9)	18282	(18)	521	(28)	1827
		(19)	558	(29)	3859

TABLE 21

Values of x^*_w-estimate of linear age effect

$x^*_w(ijk)$

Breathlessness			Yes, j = 1		No, j = 2	
			Yes	No	Yes	No
Wheeze			k = 1	k = 2	k = 1	k = 2
	1	20 - 24	10.212	5.788	93.788	1842.212
	2	25 - 29	21.210	10.790	106.789	1652.210
Age	3	30 - 34	52.491	20.509	178.509	1861.490
groups	4	35 - 39	121.388	47.612	256.612	2357.387
(years) i =	5	40 - 44	169.291	53.709	272.710	1778.289
	6	45 - 49	274.769	82.231	318.231	1717.768
	7	50 - 54	393.339	127.661	255.661	1313.339
	8	55 - 59	418.790	139.211	212.211	979.789
	9	60 - 64	365.509	112.491	138.490	519.510

TABLE 22

Analysis of Information

Component due to	Information	D.F.
x(ij.),x(i.k),x(.jk)	$2I(x:x_2^*) = 26.673$	8
x(ij.),x(i.k),x(.jk),$\sum_1^9 (9 - i)x(i11)$	$2I(x_w^*:x_2^*) = 19.873$	1
	$2I(x:x_w^*) = 6.800$	7

The log-odds or logit representations for the estimate x_w^* are

$$\ln(x_w^*(i1k)/x_w^*(i2k)) = \tau_1^B + \tau_{i1}^{AB} + \tau_{11}^{BW} + (9 - i)(2 - k)\tau^{ABW},$$

$$\ln(x_w^*(ij1)/x_w^*(ij2)) = \tau_1^W + \tau_{i1}^{AW} + \tau_{11}^{BW} + (9 - i)(2 - j)\tau^{ABW},$$

where the values of the natural parameters are

1. $\tau_1^B = -1.5300$
2. $\tau_1^W = -1.3221$
3. $\tau_{11}^{AB} = -4.2330$
4. $\tau_{21}^{AB} = -3.5013$
5. $\tau_{31}^{AB} = -2.9783$
6. $\tau_{41}^{AB} = -2.3722$
7. $\tau_{51}^{AB} = -1.9698$
8. $\tau_{61}^{AB} = -1.5092$
9. $\tau_{71}^{AB} = -0.8009$
10. $\tau_{81}^{AB} = -0.4213$
11. $\tau_{11}^{AW} = -1.6556$
12. $\tau_{21}^{AW} = -1.4169$
13. $\tau_{31}^{AW} = -1.0224$
14. $\tau_{41}^{AW} = -0.8957$
15. $\tau_{51}^{AW} = -0.5529$
16. $\tau_{61}^{AW} = -0.3639$
17. $\tau_{71}^{AW} = -0.3144$
18. $\tau_{81}^{AW} = -0.2077$
19. $\tau_{11}^{BW} = -2.5005$
20. $\tau^{ABW} = 0.1306$

It resulted that the variance of τ^{ABW} is 0.00087. It is found that $X^2 = (0.1306)^2/0.00087 = 19.605$, as a chi-square with one degree of freedom, is a quadratic approximation to $2I(x_w^*:x_2^*)$. Both test the null hypothesis that $\tau^{ABW} = 0$. A 95% confidence interval for the natural parameter τ^{ABW} is given by

$$0.1306 \pm 1.96(0.00087)^{1/2} \text{ or } 0.0728,\ 0.1884.$$

These results are in agreement with those obtained by a different approach by Plackett (1974, 95).

A 95% confidence interval for the linear moment parameter $\sum_{i=1}^{9} (9 - i)x_w^*(i11) = \sum_{i=1}^{9} (9 - i)x(i11)$ may be obtained by $3859 \pm 1.96(1149.425)^{1/2}$, where $1149.425 = 1/0.00087$, and yields 3792.550, 3925.450. We summarize below values obtained for MDI estimates fitting the two-way marginals of the original data but varying values of the linear moment parameter.

Linear Moment parameter	τ^{ABW}	Information
3925.45	0.1892	10.720
3859	0.1306	6.802
3792.55	0.0734	10.592
3791.99	0.0728	10.664

Note that the information statistics for the MDI estimates corresponding to the extremes of the confidence intervals are approximately 6.802 + 3.841 = 10.643. The values corresponding to $\tau^{ABW} = 0.0728$ were obtained by iteratively fitting the two-way marginals of the original data but starting with an initial distribution computed so that it had the required value for τ^{ABW}. Such a distribution, say y(ijk) may be obtained by starting with the nine two-way tables, 1, 1, 1, $\exp(0.0728(9 - i))$, $i = 1, \ldots, 9$, and then scaling each age total to the value x(i..).

EXAMPLE 8

Injury producing accidents. This example deals with the same data as Chapter 7 Section IV but from an ICP approach rather than the ECP approach previously used. It illustrates the use of OUTLIER when there is second-order interaction, as well as further application of the Newton-Raphson type procedure. The use of a constraint on a single cell is illustrated, as well as the use of a distribution other than the uniform as the initial distribution in an ICP. In the latter case we see which natural parameters (taus) are affected.

INJURY PRODUCING ACCIDENTS

In Chapter 7 we considered Schotz's data as a 4 x 2 x 2 contingency table with respect to the hypothesis of no linear second-order interaction. We now reconsider the data in terms of the concept of no second-order interaction as originally defined by Bartlett, that is, multiplicative or loglinear. We shall treat the data as a 2 x 2 x 4 table and denote the occurrences in the 2 x 2 x 4 Table 1 by $x(jk\ell)$ with the notation

Variable	Index	1	2	3	4
Accident type	j	Rollover	Non-rollover		
Driver group	k	Lone driver	Injured driver with passengers		
Accident severity	ℓ	Minor	Moderate	Moderately severe	Severe to extreme

We shall first summarize the results of the analysis. To obtain the MDI estimate under the classical model of no second-order interaction we of course fit the two-way marginals $x(jk.), x(j.\ell), x(.k\ell)$. Denoting the MDI estimate by $x_2^*(jk\ell)$ it is found that $2I(x:x_2^*) = 10.489$, 3 D.F., which is significant at the .025 level. The basic model x_a^* fitting the marginals $x(jk.), x(..\ell)$ where $x_a^*(jk\ell) = x(jk.)x(..\ell)/n$, is the MDI estimate under a null hypothesis of homogeneity of the four-cell multinomials of the

TABLE 1

Driver Group (k)	Accident Type (j)	Accident Severity (ℓ)			
		Minor	Moderate	Moderately Severe	Severe to Extreme
Lone Driver	Rollover	21	567	1356	644
	Non-rollover	996	5454	2773	1256
Injured Driver with Passengers	Rollover	18	553	1734	869
	Non-rollover	679	4561	2516	1092

dependent variable accident severity over the four combinations of the explanatory variables accident type and driver group. It is found that $2I(x:x_a^*) = 3243.591$, 9 D.F. We summarize these results in Analysis of Information Table 2.

TABLE 2

Analysis of Information

Component due to	Information	D.F.
a) $x(jk.), x(..\ell)$	$2I(x:x_a^*) = 3243.591$	9
$x(jk.), x(j.\ell), x(.k\ell)$ (no second-order interaction)	$2I(x_2^*:x_a^*) = 3233.102$	6
	$2I(x:x_2^*) = 10.489$	3

$$2I(x_2^*:x_a^*)/2I(x:x_a^*) = 3233.102/3243.591 = 0.997.$$

We recall that in Chapter 7, for the null hypothesis of no linear second-order interaction, $2I(x^*:x) = 19.703$, 3 D.F., leading to rejection of that null hypothesis.

Since the MDI estimate x_2^* accounts for almost all the variation unexplained by the base MDI estimate x_a^*, it would seem reasonable to use it for further analysis, despite the indicated significance level, since $n = 25089$ is large. However for its methodologic interest we shall seek an MDI estimate to account for the presence of second-order interaction.

The computer output for the MDI estimate x_2^* showed much larger OUTLIER values for the cells 112 and 122 compared to the other cells. If we seek the MDI estimate x^* fitting all two-way marginals with the additional constraint $x^*(112) = x(112)$, then we can proceed by considering the 2 x 2 x 3 table omitting all the observations corresponding to $\ell = 2$, Moderate accident severity, because of the relations

$$x(112) + x(212) = x(.12),\ x(112) + x(122) = x(1.2),$$

$$x(112) + x(122) + x(212) + x(222) = x(..2).$$

We remark that the computer output for the 2 x 2 x 3 table, and for the 2 x 2 x 4 table fitting the two-way marginals with the additional constraint $x^*(112) = x(112)$, yielded the same estimates, natural parameters (taus), and covariance matrix of the taus for the cells with the index $\ell = 1,3,4$.

We summarize results in Analysis of Information Table 3.

TABLE 3

Analysis of Information

Component due to	Information	D.F.
a) $x(jk.), x(..\ell)$	$2I(x:x_a^*) = 3243.591$	9
$x(jk.), x(j.\ell), x(.k\ell)$ (no second-order interaction)	$2I(x_2^*:x_a^*) = 3233.102$	6
	$2I(x:x_2^*) = 10.489$	3
$x(jk.), x(j.\ell), x(.k\ell)$ and $x^*(112) = x(112)$	$2I(x^*:x_2^*) = 8.869$	1
	$2I(x:x^*) = 1.620$	2
With cells $x(jk2)$ omitted		
b) $x(jk.), x(..\ell)$	$2I(x:x_b^*) = 1204.535$	6
$x(jk.), x(j.\ell), x(.k\ell)$	$2I(x^*:x_b^*) = 1202.925$	4
	$2I(x:x^*) = 1.610$	2

In Table 4 are listed the original observations, as well as x_a^*, x_b^*, x_2^*, x^* with OUTLIER omitted, x^* with constraint $x^*(112) = x(112)$ and also the estimate using "ridits" from Chapter 7. The

TABLE 4

jkℓ	x(jkℓ)	$x_a^*(jk\ell)$	$x_b^*(jk\ell)$	$x_2^*(jk\ell)$
111	21	176.804	248.244	20.18
112	567	1148.606		526.33
113	1356	864.317	1213.556	1369.31
114	644	398.273	559.200	672.18
121	18	216.837	321.943	18.82
122	553	1408.685		593.67
123	1734	1060.024	1573.840	1720.69
124	869	488.454	725.217	840.82
211	996	715.891	617.231	996.82
212	5454	4650.790		5494.67
213	2773	3499.683	3017.377	2759.69
214	1256	1612.636	1390.392	1227.82
221	679	604.467	526.581	678.18
222	4561	3926.919		4520.34
223	2516	2954.976	2574.228	2529.31
224	1092	1361.638	1186.191	1120.18

Omit OUTLIER x*(jkℓ)	x*(jkℓ) x*(112) = x(112)	"Ridit"
19.64	19.64	19.95
	567.00	551.12
1342.05	1342.05	1359.58
659.31	659.31	657.35
19.36	19.36	18.79
	553.00	566.72
1747.95	1747.95	1732.76
853.69	853.69	855.73
997.36	997.36	1004.56
	5454.00	5469.99
2786.96	2786.96	2759.92
1240.69	1240.69	1244.54
677.64	677.64	671.88
	4561.00	4543.51
2502.04	2502.04	2529.14
1107.32	1107.32	1103.47

k-sample version of the Newton-Raphson type iteration described in Chapter 5 was used. The initial distribution for the iteration was taken to be x_a^* rather than the uniform, to illustrate the statements in Chapter 3 about possible choices of the initial distribution for the iterative procedure, and the computation of the natural parameters (taus) in such situations.. In this case then, the representation for the relative log-odds is

$$\ell n(x^*(jk1)/x^*(jk4)) = \tau_1^{\ell} + \tau_{11}^{j\ell} + \tau_{11}^{k\ell}$$

(1)
$$\ell n(x^*(jk2)/x^*(jk4)) = \tau_2^{\ell} + \tau_{12}^{j\ell} + \tau_{12}^{k\ell} + \tau_{112}^{jk\ell}$$

$$\ell n(x^*(jk3)/x^*(jk4)) = \tau_3^{\ell} + \tau_{13}^{j\ell} + \tau_{13}^{k\ell}.$$

Since the initial distribution x_a^* has the representation for the relative log-odds

$$\ell n(x_a^*(jk1)/x_a^*(jk4)) = \tau_1^{\ell}, \quad \ell n(x_a^*(jk2)/x_a^*(jk4)) = \tau_2^{\ell},$$

$$\ell n(x_a^*(jk3)/x_a^*(jk4)) = \tau_3^{\ell},$$

the values of the natural parameters τ_1^{ℓ}, τ_2^{ℓ}, τ_3^{ℓ}, given in the output are

$$0.321022 = \ell n(x^*(221)/x^*(224)) - \ell n(x_a^*(221)/x_a^*(224)),$$

$$0.356437 = \ell n(x^*(222)/x^*(224)) - \ell n(x_a^*(222)/x_a^*(224)),$$

$$0.040369 = \ell n(x^*(223)/x^*(224)) - \ell n(x_a^*(223)/x_a^*(224)),$$

whereas the second-order and third-order natural parameters in (1) above are independent of x_a^*. Thus the values of the natural parameters in the representation in (1) are:

$\tau_1^{\ell} = -0.491076$ $\quad\tau_{11}^{j\ell} = -3.295304$ $\quad\tau_{11}^{k\ell} = 0.272769$

$\tau_2^{\ell} = 1.415604$ $\quad\tau_{12}^{j\ell} = -1.849810$ $\quad\tau_{12}^{k\ell} = 0.065082$

$\tau_3^{\ell} = 0.815172$ $\quad\tau_{13}^{j\ell} = -0.098535$ $\quad\tau_{13}^{k\ell} = -0.005885$

$$\tau_{112}^{jk\ell} = 0.218284$$

The variance of $\tau_{112}^{jk\ell}$ is 0.005366. Since the value of $\tau_{112}^{jk\ell}$ is zero for x_2^* it is found that $X^2 = (0.218284)^2/0.005366 = 8.880$ is a quadratic approximation to $2I(x^*:x_2^*) = 8.869$. Note also that $(x^*(112) - x_2^*(112))^2(0.005366) = (567.00 - 526.33)^2(.005366) =$ 8.876 is the moment quadratic approximation to $2I(x^*:x_2^*)$.

The associations, measured by the logarithm of the cross-product ratio, between accident type and accident severity are:

$$\ell n(x^*(1k1)/x^*(1k4)) - \ell n(x^*(2k1)/x^*(2k4)) = \tau_{11}^{j\ell} = -3.2953,$$

$k = 1,2,$

$$\ell n(x^*(112)/x^*(114)) - \ell n(x^*(212)/x^*(214)) = \tau_{12}^{j\ell} + \tau_{112}^{jk\ell} = -1.6315,$$

$$\ell n(x^*(122)/x^*(124)) - \ell n(x^*(222)/x^*(224)) = \tau_{12}^{j\ell} = -1.8498,$$

$$\ell n(x^*(1k3)/x^*(1k4)) - \ell n(x^*(2k3)/x^*(2k4)) = \tau_{13}^{j\ell} = -0.0985,$$

$k = 1,2.$

The associations, measured by the logarithm of the cross-product ratio, between driver group and accident severity are:

$$\ell n(x^*(j11)/x^*(j14)) - \ell n(x^*(j21)/x^*(j24)) = \tau_{11}^{k\ell} = 0.2728,\ j = 1,2,$$

$$\ell n(x^*(112)/x^*(114)) - \ell n(x^*(122)/x^*(124)) = \tau_{12}^{k\ell} + \tau_{112}^{jk\ell} = 0.2834,$$

$$\ell n(x^*(212)/x^*(214)) - \ell n(x^*(222)/x^*(224)) = \tau_{12}^{k\ell} = 0.0651,$$

$$\ell n(x^*(j13)/x^*(j14)) - \ell n(x^*(j23)/x^*(j24)) = \tau_{13}^{k\ell} = -0.0059,$$

$j = 1,2.$

We see from the above that the relative log-odds for accident severity (relative to severe to extreme) are larger for accident type non-rollover than for rollover. The relative log-odds are

TABLE 5

B-matrix, Two-way Marginals

j	1	1	1	1	1	1	1	1	2	2	2	2	2	2	2	2
k	1	1	1	1	2	2	2	2	1	1	1	1	2	2	2	2
ℓ	1	2	3	4	1	2	3	4	1	2	3	4	1	2	3	4
	1	1	1	1	0	0	0	0	0	0	0	0	0	0	0	0
	0	0	0	0	1	1	1	1	0	0	0	0	0	0	0	0
	0	0	0	0	0	0	0	0	1	1	1	1	0	0	0	0
	0	0	0	0	0	0	0	0	0	0	0	0	1	1	1	1
	1	0	0	0	1	0	0	0	1	0	0	0	1	0	0	0
	0	1	0	0	0	1	0	0	0	1	0	0	0	1	0	0
	0	0	1	0	0	0	1	0	0	0	1	0	0	0	1	0
	1	0	0	0	1	0	0	0	0	0	0	0	0	0	0	0
	0	1	0	0	0	1	0	0	0	0	0	0	0	0	0	0
	0	0	1	0	0	0	1	0	0	0	0	0	0	0	0	0
	1	0	0	0	0	0	0	0	1	0	0	0	0	0	0	0
	0	1	0	0	0	0	0	0	0	1	0	0	0	0	0	0
	0	0	1	0	0	0	0	0	0	0	1	0	0	0	0	0

TABLE 6

B-matrix, Omit Outlier

j	1	1	1	1	1	1	2	2	2	2	2	2
k	1	1	1	2	2	2	1	1	1	2	2	2
ℓ	1	3	4	1	3	4	1	3	4	1	3	4
	1	1	1	0	0	0	0	0	0	0	0	0
	0	0	0	1	1	1	0	0	0	0	0	0
	0	0	0	0	0	0	1	1	1	0	0	0
	0	0	0	0	0	0	0	0	0	1	1	1
	1	0	0	1	0	0	1	0	0	1	0	0
	0	1	0	0	1	0	0	1	0	0	1	0
	1	0	0	1	0	0	0	0	0	0	0	0
	0	1	0	0	1	0	0	0	0	0	0	0
	1	0	0	0	0	0	1	0	0	0	0	0
	0	1	0	0	0	0	0	1	0	0	0	0

larger for driver group lone driver than for injured driver with passengers.

The appropriate B-matrices for the k-sample Newton-Raphson type iteration for fitting the two-way marginals and the data less the outliers are given in Tables 5 and 6 respectively.

For the MDI estimate $x^*(ijk)$ fitting the two-way marginals for the 2 x 2 x 4 table but such that $x^*(112) = x(112)$ the B-matrix will be the same as that for $x_2^*(ijk)$ (fitting the two-way marginals) except that the following 14th row must be added to the B-matrix of Table 5

0 1 0 0 0 0 0 0 0 0 0 0 0 0 0 0

BIBLIOGRAPHY

The bibliography lists publications, reports, etc., primarily dealing with the analysis of contingency tables. Items are listed by year starting with the most recent listing. Additional references to related topics may be found in the bibliographies contained in the books by Bishop, Fienberg, Holland (1975), D. R. Cox (1970), and H. O. Lancaster (1969). The bibliography depends in large part on compilations prepared by Dr. Marvin A. Kastenbaum and Dr. H. H. Ku. Permission to use their results is gratefully acknowledged. See also Ruth Killion and Douglas A. Zahn, "A Bibliography of Contingency Table Literature: 1900-1974," Int. Stat. Rev., 44 (1), 71-112 (1976). We make no claim that all items that should have been included are contained herein, and we express our regrets to authors of items omitted.

1976/5

KULLBACK, S., and J. CORNFIELD, An information theoretic contingency table analysis of the Dorn study of smoking and mortality, Computers and Biomedical Research 9, 409-437 (1976).

BISHOP, Y.M.M., S. FIENBERG, P. W. HOLLAND, Discrete Multivariate Analysis, The MIT Press, Cambridge, Massachusetts (1975).

GODAMBE, A. V. and W. L. HARKNESS, Normal approximation to the distribution of a cell entry in a 2 x 2 x 2 contingency table, Communications in Statistics, 4 (8), 699-709 (1975).

GOODMAN, L. A., On the relationship between two statistics pertaining to tests of three-factor interaction in contingency tables, Journal American Statistical Association, 70(351), 624-5 (1975).

KEEGEL, J. C., Several numerical procedures in regression and parameter estimation in contingency tables. Ph.D. dissertation, The Graduate School of Arts and Sciences, The George Washington University, February 1975.

KULLBACK, S., The information in contingency tables. Proceedings of the twentieth conference on The Design of Experiments in Army Research, Development and Testing, ARO Report 75-2, 49-53 (1975).

KULLBACK, S., and M. FISHER, Multivariate logit analysis, Biometrische Zeitschrift 17(3), 139-146 (1975).

KULLBACK, S., and P. N. REEVES, Analysis of interactions between categorical variables, Biometrische Zeitschrift, 17(1), 3-12 (1975).

MANTEL, N., and J. L. FLEISS, The equivalence of the generalized McNemar tests for marginal homogeneity in 2^3 and 3^2 tables, Biometrics, 31(3), 727-29 (1975).

SCHEUREN, F. J., and H. L. OH, A data analysis approach to fitting square tables, Communications in Statistics, 4(7), 595-615 (1975).

1974

BENNETT, B. M., and C. KANESHIRO, On the small-sample properties of the Mantel-Haenszel test for relative risk, Biometrika, 61, 233-236 (1974).

BERKSON, J., and B. N. NAGNUR, A note on the minimum χ_1^2 estimate and a L.A.M.S.T. χ^2 in the "No Interaction" problem, Journal Am. Statist. Assn., 69, 1038-1040 (1974).

BROWN, M. B., Identification of the sources of significance in two-way contingency tables, Applied Statistics, 23(3), 405-413 (1974).

DARROCH, J. N., Multiplicative and additive interaction in contingency tables, Biometrika 61(2), 207-214 (1974).

FISHER, M. R., User's guide to CONTABMOD, Statistics Department, The George Washington University, 1974. (Now at National Heart and Lung Institute, Bethesda, Md. 20014)

GAIL, M., Value systems for comparing two independent multinomial trials, Biometrika 61(1), 91-100 (1974).

HABERMAN, S. J., The Analysis of Frequency Data. The University of Chicago Press, 1974.

KASTENBAUM, M. A., Analysis of categorical data: some wellknown analogues and some new concepts, Communications in Statistics 3(5), 401-417 (1974).

KU, H. H., and S. KULLBACK, Loglinear models in contingency table analysis, The American Statistician 28(4), 115-122 (1974).

KULLBACK, S., The Information in Contingency Tables, Final Technical Report, September 1974, U.S. Army Research Office, Durham, N.C., Grant No. DAHCO 4-74-G-0164.

NELDER, J. A., Log-linear models for contingency tables: A generalization of classical least squares. Applied Statistics 23(3), 323-329 (1974).

PATIL, K. D., Interaction test for three-dimensional contingency tables, Journal Am. Statist. Assn. 69(345), 164-168 (1974).

PLACKETT, R. L., The Analysis of Categorical Data, Griffin's Statistical Monographs and Courses, No. 35. Griffin, London, 1974.

SIMON, G., Alternative analyses for the singly-ordered contingency table, Journal Am. Statist. Assn. 69, 971-976 (1974).

SUGIURA, N., Maximum likelihood estimates for the logit model and the iterative scaling method, Communications in Statistics 3(10), 985-993 (1974).

SUGIURA, N., and M. OTAKE, An extension of the Mantel-Haenszel procedure to k 2 x c contingency tables and the relation to the logit model, Communications in Statistics 3(9), 829-842 (1974).

1973

GOKHALE, D. V., Approximating discrete distributions with applications, Journal Am. Statist. Assn. 68(344), 1009-1012 (1973).

GOODMAN, L. A., Guided and unguided methods for selecting models for a set of T multidimensional contingency tables, Journal Am. Statist. Assn. 68, 165-175 (1973).

KULLBACK, S., Estimating and testing interaction parameters in the log-linear model, Biometrische Zeitschrift 15, 371-388 (1973).

KULLBACK, S., and M. FISHER, Partitioning second-order interaction in three-way contingency tables, Journal Royal Statist. Soc., Series C (Applied Statistics) 22, 172-184 (1973).

MANTEL, N., and C. BROWN, A logistic reanalysis of Ashford and Sowden's data on respiratory symptoms in British coal miners, Biometrics 29(4), 649-665 (1973).

NERLOVE, M., and S. J. PRESS, Univariate and Multivariate Log-Linear and Logistic Models. R-1306-EDA/NIH December 1973, Rand, Santa Monica, Cal. 90406.

SIMON, G., Additivity of information in exponential family probability laws, Journal Am. Statist. Assn. 68, 478-482 (1973).

SUGIURA, N., and M. OTAKE, Approximate distribution of the maximum of c - 1 χ^2-statistics (2 x 2) derived from 2 x c contingency tables, Communications in Statistics 1(1), 9-16 (1973).

1972

ADAM, J., and H. ENKE, Analyse mehrdimensionaler kontingenztafeln mit hilfe des informationsmasses von Kullback, Biometrische Zeitschrift 14(5), 305-323 (1972).

BERKSON, J., Minimum discrimination information, the "no interaction" problem, and the logistic function, Biometrics 28(2), 443-468 (1972).

BRUNDEN, M. N., The analysis of non-independent 2 x 2 tables from 2 x c tables using rank sums, Biometrics 28(2), 603-606 (1972).

CAUSEY, B. D., Sensitivity of raked contingency table totals to changes in problem conditions, Ann. Math. Statist., 43(2), 656-658 (1972).

COX, D. R., The analysis of multivariate binary data, Appl. Statist., 21(2), 113-120 (1972).

DARROCH, J. N., and D. RATCLIFF, Generalized iterative scaling for log-linear models, Ann. Math. Statist., 43(5), 1470-1480 (1972).

FIENBERG, S. E., The analysis of incomplete multiway contingency tables, Biometrics, 28(1), 177-202 (1972).

FISHER, M. R., An application of minimum discrimination information estimation, Ph.D. dissertation, The Graduate School of Arts and Sciences, The George Washington University, September 1972.

GAIL, M. H., Mixed quasi-independence models for categorical data, Biometrics, 28(3), 703-712 (1972).

GART, J. J., Interaction tests for 2 x s x t contingency tables, Biometrika, 59(2), 309-316 (1972).

GOKHALE, D. V., Analysis of log-linear models, J. Roy. Statist. Soc. Ser. B, 34(3), 371-376 (1972).

GOODMAN, L. A., and W. H. KRUSKAL, Measures of association for cross-classifications, IV: simplification of asymptotic variances, J. Amer. Statist. Assn., 67, 415-421 (1972).

GRIZZLE, J. E., and O. D. WILLIAMS, Log-linear models and tests of independence for contingency tables, Biometrics, 28(1), 137-156 (1972).

GRIZZLE, J. E., and O. D. WILLIAMS, Contingency tables having ordered response categories, J. Amer. Statist. Assoc. 67, 55-63 (1972).

IRELAND, C. T., Sequential cell deletion in contingency tables, Statistics Department, The George Washington University, 1972.

JACOBS, S. E., P. N. REEVES, and G. L. HAMMON, Your guide to surveys of hospital computer usage, Hospital Financial Management 26(9), 5-13 (1972).

KOCH, G. G., P. B. IMREY, and D. W. REINFURT, Linear model analysis of categorical data with incomplete response vectors, Biometrics, 28(3), 663-692 (1972).

MARTIN, D. C., and R. A. BRADLEY, Probability models, estimation, and classification for multivariate dichotomous populations, Biometrics, 28, 203-221 (1972).

NATHAN, G., Asymptotic power of tests for independence in contingency tables from stratified samples, J. Amer. Statist. Assn., 67, 917-920 (1972).

VICTOR, N., Zur klassifizierung mehrdimensionaler kontingenztafeln, Biometrics, 28(2), 427-442 (1972).

1971

ALTHAM, P. M. E., Exact Bayesian analysis of the intraclass 2 x 2 table, Biometrika, 58(3), 679-680 (1971).

ALTHAM, P. M. E., The analysis of matched proportions, Biometrika, 58(3), 561-576 (1971).

BELLE, G. V., and R. G. CORNELL, Strengthening tests of symmetry in contingency tables, Biometrics, 27, 1074-1078 (1971).

BISHOP, Y. M. M., Effects of collapsing multidimensional contingency tables, Biometrics, 27, 545-562 (1971).

COHEN, J. E., Estimation and interaction in a censored 2 x 2 x 2 contingency table, Biometrics, 27, 379-386 (1971).

DEMPSTER, A. P., An overview of multivariate data analysis, Journal Multivariate Analysis, 1, 316-347 (1971).

FRYER, J. G., On the homogeneity of the marginal distributions of a multidimensional contingency table, J. Roy. Statist. Soc. Ser. A, 134, 368-371 (1971).

GART, J. J., On the ordering of contingency tables for significance tests, Technometrics 13, 910-911 (1971).

GART, J. J., The comparison of proportions: a review of significance tests, confidence intervals, and adjustments for stratification, Rev. Inst. Internat. Statist. 29, 148-169 (1971).

GOKHALE, D. V., An iterative procedure for analysing log-linear models, Biometrics 27, 681-687 (1971).

GOODMAN, L. A., Partitioning of chi-square, analysis of marginal contingency tables, and estimation of expected frequencies in multi-dimensional contingency tables, J. Amer. Statist. Assn. 66, 339-344 (1971).

GOODMAN, L. A., Some multiplicative models for the analysis of cross-classified data, Proc. 6th Berkeley Symp., Berkeley and Los Angeles, University of California Press, 1971.

GOODMAN, L. A., The analysis of multidimensional contingency tables: stepwise procedures and direct estimation methods for building models for multiple classifications, Technometrics 13, 33-61 (1971).

GRIZZLE, J. E., Multivariate logit analysis, Biometrics 27, 1057-1062 (1971).

IRELAND, C. T., A computer program for analyzing contingency tables. (Latest version is CONTAB III) Statistics Department, The George Washington University, 1971.

JOHNSON, W. D., and G. G. KOCH, A note on the weighted least squares analysis of the Ries-Smith contingency table data, Technometrics 13, 438-447 (1971).

KOCH, G. G., P. B. IMREY, and D. W. REINFURT, Linear model analysis of categorical data with incomplete response vectors, Institute of Statistics Mimeo Series No. 790, University of North Carolina, 1971.

KOCH, G. G., W. D. JOHNSON and H. D. TOLLEY, An application of linear models to analyze categorical data pertaining to the relationship between survival and extent of disease, Institute of Statistics Mimeo Series No. 770, University of North Carolina, 1971.

KOCH, G. G., and D. W. REINFURT, The analysis of categorical data from mixed models, Biometrics 27, 157-173 (1971).

KU, H. H., Analysis of information - an alternative approach to the detection of a correlation between the sexes of adjacent sibs in human families, Biometrics 27, 175-182 (1971).

KU, H. H., R. VARNER and S. KULLBACK, On the analysis of multidimensional contingency tables, J. Amer. Statist. Assoc. 66, 55-64 (1971).

KULLBACK, S., Marginal homogeneity of multidimensional contingency tables, Ann. Math. Statist. 42, 594-606 (1971).

KULLBACK, S., The homogeneity of the sex ratio of adjacent sibs in human families, Biometrics 27, 452-457 (1971).

NAM, J., On two tests for comparing matched proportions, Biometrics 27, 945-959 (1971).

PEACOCK, P. B., The non-comparability of relative risks from different studies, Biometrics 27, 903-907 (1971).

PERITZ, E., Estimating the ratio of two marginal probabilities in a contingency table, Biometrics 27, 223-225 (1971).

RATCLIFF, D., Topics on independence and correlation for bounded sum variables, Ph.D. thesis, School of Mathematical Sciences, the Flinders University of South Australia, June 1971.

SIMON, G. A., Information distances and exponential families, with applications to contingency tables. Technical Report No. 32, November 26, 1971, Department of Statistics, Stanford University.

THOMAS, D. G., Exact confidence limits for the odds ratio in a 2 x 2 table, Appl. Statist. 20, 105-110 (1971).

YASAIMAIBODI (YASSAEE), H., On comparison of various estimators and their associated statistics in r x c and r x c x 2 contingency tables, Ph.D. dissertation, The George Washington University, 1971.

ZELEN, M., The analysis of several 2 x 2 contingency tables, Biometrika 58, 129-137 (1971).

1970

ALTHAM, P. M. E., The measurement of association of rows and columns for an r x s contingency table, J. Roy. Statist. Soc. Ser. B **32**, 63-73 (1970).

ASHFORD, J. R., D. C. MORGAN, S. RAE, and R. R. SOWDEN, Respiratory symptoms in British coal miners, American Review of Respiratory Disease, **102**, 370-381 (1970).

ASHFORD, J. R., and R. D. SOWDEN, Multivariate probit analysis, Biometrics **26**, 535-546 (1970).

BHAPKAR, V. P., Categorical data analysis of some multivariate tests, Essays in Probability and Statistics (R. C. Bose et al., eds.), The University of North Carolina Press, 85-110, 1970.

CAMPBELL, L. L., Equivalence of Gauss's principle and minimum discrimination information estimation of probabilities, Ann. Math. Statist., **41**, 1011-1015 (1970).

COX, D. R., The Analysis of Binary Data, Methuen & Co., Ltd., London, 1970.

CRADDOCK, J. M., and C. R. FLOOD, The distribution of the chi-square statistic in small contingency tables, Appl. Statist., **19**, 173-181 (1970).

FIENBERG, S. E., Quasi-independence and maximum likelihood estimation in incomplete contingency tables, J. Amer. Statist. Assoc. **65**(332), 1610-1616 (1970).

FIENBERG, S. E., An iterative procedure for estimation in contingency tables, Ann. Math. Statist. **41**, 907-917 (1970).

FIENBERG, S. E., The analysis of multidimensional contingency tables, Ecology **51**(2), 419-433 (1970).

FIENBERG, S. E., and J. P. GILBERT, Geometry of a two by two contingency table, J. Amer. Statist. Assoc. **65**, 694-701 (1970).

FIENBERG, S. E., and P. W. HOLLAND, Methods for eliminating zero counts in contingency tables, Random Counts in Scientific Work (G. P. Patil, ed.), The Pennsylvania State University Press, 1970.

GOOD, I. J., T. N. GOVER, and G. J. MITCHELL, Exact distributions for chi-squared and for the likelihood-ratio statistic for the equiprobable multinomial distribution, J. Amer. Statist. Assoc. **65**, 267-283 (1970).

GOODMAN, L. A., The multivariate analysis of qualitative data: interaction among multiple classifications, J. Amer. Statist. Assoc. **65**, 226-256 (1970).

HABERMAN, S., The general log-linear model, Ph.D. Dissertation, University of Chicago, September 1970.

KASTENBAUM, M. A., A review of contingency tables, Essays in Probability and Statistics (R. C. Bose et al., eds.). The University of North Carolina Press, 1970, 407-438.

KULLBACK, S., Various applications of minimum discrimination information estimation, particularly to problems of contingency table analysis, Proceedings of the Meeting on Information Measures, University of Waterloo, Ontario, Canada, April 10-14, 1970, I-33-I-66.

KULLBACK, S., Minimum discrimination information estimation and application, Proceedings of the Sixteenth Conference on the Design of Experiments in Army Research, Development and Testing, 21 October 1970, ARO-D Report 71-3, 1-38 Proceedings of the Conference.

MANTEL, N., Incomplete contingency tables, Biometrics **26**, 291-304 (1970).

MOLK, Y., On estimation of probabilities in contingency tables with restrictions on marginals, Ph.D. dissertation, The George Washington University, February 1970.

ODOROFF, C. L., Minimum logit chi-square estimation and maximum likelihood estimation in contingency tables, J. Amer. Statist. Assoc. **65**(332), 1617-1631 (1970).

WAGNER, S. S., The maximum-likelihood estimate for contingency tables with zero diagonal, J. Amer. Statist. Assoc. 65(331), 1362-1383 (1970).

1969

ALTHAM, P. M. E., Exact Bayesian analysis of a 2 x 2 contingency table and Fisher's "exact" significance test, J. Roy. Statist. Soc. Ser. B. 31, 261-269 (1969).

ARGENTIERO, P. D., χ-squared statistic for goodness of fit test, its derivation and tables, NASA Technical Report, TR-R-313, 1969.

BISHOP, Y. M. M., Full contingency tables, logits, and split contingency tables, Biometrics 25, 383-400 (1969).

BISHOP, Y. M. M., and S. E. FIENBERG, Incomplete two-dimensional contingency tables, Biometrics 25, 119-128 (1969).

DEMPSTER, A. P., Some theory related to fitting exponential models, Research Report S-4, Department of Statistics, Harvard University, 1969.

FIENBERG, S. E., Preliminary graphical analysis and quasi-independence for two-way contingency tables, Appl. Statist. 18, 153-168 (1969).

GOODMAN, L. A., On partition chi-squared and detecting partial association in the three-way contingency tables, J. Roy. Statist. Soc. Ser. B. 31, 486-498 (1969).

GRIZZLE, J. E., C. F. STARMER, and G. G. KOCH, Analysis of categorical data by linear models, Biometrics 25, 489-504 (1969).

HEALY, M. J. R., Exact tests of significance in contingency tables, Technometrics 11, 393-395 (1969).

IRELAND, C. T., H. H. KU, and S. KULLBACK, Symmetry and marginal homogeneity of an r x r contingency table, J. Amer. Statist. Assoc. 64, 1323-1341 (1969).

KOCH, G. G., The effect of non-sampling errors on measures of association in 2 x 2 contingency tables, J. Amer. Statist. Assoc. 64, 852-863 (1969).

KU, H. H., and S. KULLBACK, Analysis of multidimensional contingency tables: an information theoretical approach. Contributed papers, 37th Session of the International Statistical Institute, 1969, 156-158.

KU, H. H., and S. KULLBACK, Approximating discrete probability distributions, IEEE Trans. on Information Theory, IT-15, 444-447 (1969).

LANCASTER, H. O., Contingency tables of higher dimensions, Bulletin of the International Statistical Institute, 43(1), 143-151 (1969).

LANCASTER, H. O., The Chi-Squared Distribution, Wiley, New York, 1969.

NAGNUR, B. N., LAMST and the hypotheses of no three factor interaction in contingency tables, J. Amer. Statist. Assoc. 64, 207-215 (1969).

PLACKETT, R. L., Multidimensional contingency tables. A Survey of Models and Methods, Bulletin of the International Statistical Institute. 43(1), 133-142 (1969).

1968

BENNETT, B. M., Notes on chi-squared tests for matched samples, J. Roy. Statist. Soc. Ser. B 30, 368-370 (1968).

BERKSON, J., Application of minimum logit chi-squared estimate to a problem of Grizzle with a notation on the problem of no interaction, Biometrics 24, 75-96 (1968).

BHAPKAR, V. P., On the analysis of contingency tables with a quantitative response, Biometrics 24, 329-338 (1968).

BHAPKAR, V. P., and G. G. KOCH, Hypotheses of "no interaction" in multidimensional contingency tables, Technometrics 10, 107-123 (1968).

BHAPKAR, V. P., and G. G. KOCH, On the hypotheses of "no interaction" in contingency tables, Biometrics 24, 567-594 (1968).

CENCOV, N. N., A nonsymmetric distance between probability distributions and entropy and the theorem of Pythagoras (Russian). Mat. Zametki 4, 323-332 (1968).

COX, D. R., and E. J. SNELL, A general definition of residuals (with discussion), J. Roy. Statist. Soc. Ser. B 30, 248-275 (1968).

FIENBERG, S. E., The geometry of an r x c contingency table, Ann. Math. Statist. 39, 1186-1190 (1968).

GOODMAN, L. A., The analysis of cross-classified data: independence, quasi-independence, and interactions in contingency tables with or without missing entries, J. Amer. Statist. Assoc. 63, 1091-1131 (1968).

HAMDAN, M. A., Optimum choice of classes for contingency tables, J. Amer. Statist. Assoc. 63, 291-297 (1968)

IRELAND, C. T., and S. KULLBACK, Contingency tables with given marginals, Biometrika 55, 179-188 (1968).

IRELAND, C. T., and S. KULLBACK, Minimum discrimination information estimation, Biometrics 24, 707-713 (1968).

KU, H. H., and S. KULLBACK, Interaction in multidimensional contingency tables: an information theoretic approach, J. Res. Nat. Bur. Standards Sect. B 72, 159-199 (1968).

KU, H. H., R. VARNER, and S. KULLBACK, Analysis of multidimensional contingency tables. Proceedings of the Fourteenth Conference on the Design of Experiments in Army Research, Development and Testing, ARO-D Report 69-2, 1968.

KULLBACK, S., Information Theory and Statistics, Dover Pub., Inc., New York, 1968.

KULLBACK, S., Probability densities with given marginals, Ann. Math. Statist. 39, 1236-1243 (1968).

LEYTON, M. K., Rapid calculation of exact probabilities for 2 x 3 contingency tables, Biometrics 24, 714-717 (1968).

MATHIEU, J. R., and E. LAMBERT, Un test de l'identite des marges dun tableau de correlation, C. R. Acad. Sci. Paris 267, 832-834 (1968).

MOSTELLER, F., Association and estimation in contingency tables, J. Amer. Statist. Assoc. 63, 1-28 (1968).

SLAKTER, M. J., Accuracy of an approximation to the power of the chi-square goodness of fit test with small but equal expected frequencies, J. Amer. Statist. Assoc. 63, 912-924 (1968).

SUGIURA, N., and M. OTAKE, Numerical comparison of improvised methods of testing in contingency tables with small frequencies, Ann. Inst. Statist. Math. 20, 507-517 (1968).

1967

BENNETT, B. M., Tests of hypothesis concerning matched samples, J. Roy. Statist. Soc. Ser. B 29, 468-474 (1967).

BISHOP, Y. M. M., Multidimensional contingency tables: cell estimates, Ph.D. dissertation, Harvard University, 1967.

BLOCH, D. A., and G. S. WATSON, A Bayesian study of the multinomial distribution, Ann. Math. Statist. 38, 1423-1435 (1967).

COX, D. R., and E. LAUH, A note on the graphical analysis of multidimensional contingency tables, Technometrics 9, 481-488 (1967).

GOOD, I. J., A Bayesian significance test for multinomial distributions, J. Roy. Statist. Soc. Ser. B 29, 339-431 (1967).

FERGUSON, T. S., Mathematical Statistics, Academic Press Inc., N. Y., 1967.

SNEDECOR, G. W., and W. G. COCHRAN, Statistical Methods, The Iowa State University Press, Ames, Iowa, 1967.

1966

ARMITAGE, P., The chi-square test for heterogeneity of proportions after adjustment for stratification, J. Roy. Statist. Soc. Ser. B 28, 150-163 (1966).

BHAPKAR, V. P., A note on the equivalence of two test criteria for hypotheses in categorical data, J. Amer. Statist. Assoc. 61, 228-235 (1966).

BHAPKAR, V. P., Notes on analysis of categorical data, Institute of Statistics Mimeo Series No. 477, University of North Carolina, 1966.

BHAT, B. R., and S. R. KULKARNI, LAMP test of linear and loglinear hypotheses in multinomial experiments, J. Amer. Statist. Assoc. 61, 236-245 (1966).

COX, D. R., A simple example of a comparison involving quantal data, Biometrika 53, 215-220 (1966).

CRADDOCK, J. M., Testing the significance of a 3 x 3 contingency table, The Statistician 16, 87-94 (1966).

GABRIEL, K. R., Simultaneous test procedures for multiple comparison on categorical data, J. Amer. Statist. Assoc. 61, 1081-1096 (1966).

GART, J. J., Alternative analyses of contingency tables, J. Roy. Statist. Soc. Ser. B 28, 164-179 (1966).

GOOD, I. J., How to estimate probabilities, J. Inst. Math. Appl. 2, 364-383 (1966).

KULLBACK, S., and M. A. KHAIRAT, A note on minimum discrimination information, Ann. Math. Statist. 37, 279-280 (1966).

MANTEL, N., Models for complex contingency tables and polychotomous dosage response curves, Biometrics 22, 83-95 (1966).

1965

ASANO, C., On estimating multinomial probabilities by pooling incomplete samples, Ann. Inst. Statist. Math. 17, 1-14 (1965).

BHAPKAR, V. P., and G. G. KOCH, On the hypothesis of "no interaction" in three-dimensional contingency tables. Institute of Statistics Mimeo Series No. 440, University of North Carolina, 1965.

BHAPKAR, V. P., and G. G. KOCH, Hypothesis of no "interaction" in four-dimensional contingency tables, Institute of Statistics Mimeo Series No. 449, University of North Carolina, 1965.

BHAT, B. R., and B. N. NAGNUR, Locally asymptotically most stringent tests and Lagrangian multiplier tests of linear hypotheses, Biometrika 52(3 and 4), 459-468 (1965).

BIRCH, M. W., The detection of partial association II: the general case, J. Roy. Statist. Soc. Ser. B 27, 111-124 (1965).

CAUSSINUS, H., Contribution a l'analyse statistique des tableaux de correlation, Ann. Fac. Sci. Univ. Toulouse 29, 77-182 (1965).

GOOD, I. J., The Estimation of Probabilities: An Essay on Modern Bayesian Methods. Research Monograph, 30, The MIT Press, Cambridge, Massachusetts, 1965.

KASTENBAUM, M. A., Contingency tables: a review. MRC Technical Summary Report No. 596, Mathematical Research Center, The University of Wisconsin, 1965.

KATTI, S. K., and A. N. SASTRY, Biological examples of small expected frequencies and the chi-square test, Biometrics 21, 49-54 (1965).

LANCASTER, H. O., and T. A. I. BROWN, Size of chi-squared test in the symmetrical multinomials, Austral. J. Statist. 7, 40 (1965).

LEWONTIN, R. C., and J. FELSENSTEIN, The robustness of homogeneity tests in 2 x n tables, Biometrics 21, 19-33 (1965).

MOTE, V. L., and R. L. ANDERSON, An investigation of the effect of misclassification on the properties of chi-squared tests in the analysis of categorical data, Biometrika 52, 95-109 (1965).

RADHAKRISHNA, S., Combination of results from several 2 x 2 contingency tables, Biometrics 21, 86-98 (1965).

1964

ALLISON, H. E., Computational forms for chi-square, Amer. Statist. 18(1), 17-18 (1964).

BENNETT, B. M., and E. NAKAMURA, Tables for testing significance in a 2 x 3 contingency table, Technometrics 6(4), 439-458 (1964).

BIRCH, M. W., The detection of partial association I: the 2 x 2 case, J. Roy. Statist. Soc. Ser. B 26, 313-324 (1964).

BROSS, I. D. J., Taking a covariable into account, J. Amer. Statist. Assoc. 59(307), 725-736 (1964).

CHEW, V., Application of the negative binomial distribution with probability of misclassification, Virginia Journal of Science 15(1), 34-40 (1964).

GOODMAN, L. A., Simultaneous confidence limits for cross-product ratios in contingency tables, J. Roy. Statist. Soc. Ser. B 26(1), 86-102 (1964).

GOODMAN, L. A., Simple methods for analyzing three-factor interaction in contingency tables, J. Roy. Statist. Soc. 59, 319-352 (1964).

GOODMAN, L. A., Interactions in multidimensional contingency tables, Ann. Math. Statist. 35(2), 632-646 (1964).

GOODMAN, L. A., Simultaneous confidence intervals for contrasts among multinomial populations, Ann. Math. Statist. 35(2), 716-725 (1964).

HARKNESS, W. L., and L. KATZ, Comparison of the power functions for the test of independence in 2 x 2 contingency tables, Ann. Math. Statist. 35(3), 1115-1127 (1964).

KIHLBERG, J. K., E. A. NARRAGON, and B. J. CAMPBELL, Automobile crash injury in relation to car size, Cornell Aeronautical Lab. Inc. Report, VJ-1823R11, 1964.

LINDLEY, D. V., The Bayesian analysis of contingency tables, Ann. Math. Statist. 35(4), 1622-1643 (1964).

PLACKETT, R. L., The continuity correction in 2 x 2 tables, Biometrika 21(3 and 4), 327-338 (1964).

PUTTER, J., The chi-square goodness-of-fit test for a class of cases of dependent observations, Biometrika 51, 250-252 (1964).

SOMERS, R. H., Simple measures of association for the triple dichotomy, J. Roy. Statist. Soc. Ser. A 127(3), 409-415 (1964).

TALLIS, G. M., The use of models in the analysis of some classes of contingency tables, Biometrics 24(4), 832-839 (1964).

1963

BENNETT, B. M., and E. NAKAMURA, Tables for testing significance in a 2 x 3 contingency table, Technometrics 5(4), 501-511 (1963).

BIRCH, M. W., Maximum likelihood in three-way contingency tables, J. Roy. Statist. Soc. Ser. B 25(1), 220-233 (1963).

DARROCH, J. N., and S. D. SILVEY, On testing more than one hypothesis, Ann. Math. Statist. 34(2), 555-567 (1963).

DIAMOND, E. L., The limiting power of categorical data chi-square tests analogous to normal analysis of variance, Ann. Math. Statist. 34(4), 1432-1441 (1963).

EDWARDS, A. W. F., The measure of association in a 2 x 2 table, J. Roy. Statist. Soc. Ser. A 126(1), 109-114 (1963).

FELDMAN, S. E., and E. KLINGER, Short cut calculation of the Fisher-Yates exact test, Psychometrika 28(3), 289-291 (1963).

FINNEY, D. J., R. LATSCHA, B. M. BENNETT, P. HSU, and E. S. PEARSON, Tables for testing significance in a 2 x 2 contingency table, (Supplement by B. M. Bennett and C. Horst i + 28). Camb. Univ. Press, 1963, 103 pp.

GOLD, R. A., Tests auxiliary to chi-square tests in a Markov chain, Ann. Math. Statist. 34(1), 56-74 (1963).

GOOD, I. J., Maximum entropy for hypothesis formulation, especially for multidimensional contingency tables, Ann. Math. Statist. 34(3), 911-934 (1963).

GOODMAN, L. A., On methods for comparing contingency tables, J. Roy. Statist. Soc. Ser. A 126(1), 94-108 (1963).

GOODMAN, L. A., On Plackett's test for contingency table interactions, J. Roy. Statist. Soc. Ser. B 25(1), 179-188 (1963).

GOODMAN, L. A., and W. H. KRUSKAL, Measures of association for cross classification III: approximate sampling theory, J. Amer. Statist. Assoc. 58, 310-364 (1963).

KU, H. H., A note on contingency tables involving zero frequencies and the 2I test, Technometrics 5(3), 398-400 (1963).

MANTEL, N., Chi-square tests with one degree of freedom: extensions of the Mantel-Haenszel procedure, J. Amer. Statist. Assoc. 58, 690-700 (1963).

NEWELL, D. J., Misclassification in 2 x 2 tables, Biometrics 19(1), 187-188 (1963).

OKAMATO, M., Chi-square statistic based on the pooled frequencies of several observations, Biometrika 50, 524-528 (1963).

RIES, P. N., and H. SMITH, The use of chi-square for preference testing in multidimensional problems, Chem. Eng. Prog. Symposium Series 59(42), 39-43 (1963).

WALSH, J. E., Loss in test efficiency due to misclassification for 2 x 2 tables, Biometrics 19(1), 158-162 (1963).

1962

CORNFIELD, J., Joint dependence of risk of coronary heart disease on serum cholesterol and systolic blood pressure: a discriminant function analysis, Federation Proceeding 4(II) (Suppl. No. 11), 58-61 (July-August 1962).

DALY, C., A simple test for trends in a contingency table, Biometrics 18(1), 114-119 (1962).

DARROCH, J. N., Interactions in multi-factor contingency tables, J. Roy. Statist. Soc. Ser. B 24(1), 251-263 (1962).

FISHER, SIR RONALD A., Confidence limits for a cross-product ratio, Austral. J. Statist. 4(1), 41 (1962).

GART, J. J., Approximate confidence limits for relative risks, J. Roy. Statist. Soc. Ser. B 24(2), 454-463 (1962).

GART, J. J., On the combination of relative risks, Biometrics 18(4), 601-610 (1962).

KINCAID, W. M., The combination of 2 x m contingency tables, Biometrics 18(2), 224-228 (1962).

KULLBACK, S., M. KUPPERMAN, and H. H. KU, An application of information theory to the analysis of contingency tables with a table of 2N ℓn N, N = 1(1)10,000, J. Res. Nat. Bur. Standards Sect. B 66, 217-243 (1962).

KULLBACK, S., M. KUPPERMAN, and H. H. KU, Tests for contingency tables and Markov chains, Technometrics 4(4), 573-608 (1962).

LEWIS, B. N., On the analysis of interaction in multi-dimensional contingency tables, J. Roy. Statist. Soc. Ser. A 125(1), 88-117 (1962).

PLACKETT, R. L., A note on interactions in contingency tables, J. Roy. Statist. Soc. Ser. B 24(1), 162-166 (1962).

TALLIS, G. M., The maximum likelihood estimation of correlation from contingency tables, Biometrics 18(3), 342-353 (1962).

1961

BERGER, A., On comparing intensities of association between two binary characteristics in two different populations, J. Amer. Statist. Assoc. 56, 889-908 (1961).

BHAPKAR, V. P., Some tests for categorical data, Ann. Math. Statist. 32(1), 72-83 (1961).

BILLINGSLEY, P., Statistical Inference for Markov Processes, Statistical Research Monographs, 2, The University of Chicago Press, 1961.

CLARINGBOLD, P. J., The use of orthogonal polynomials in the partition of chi-square, Austral. J. Statist. 3(2), 48-63 (1961).

FRIEDLANDER, D., A technique for estimating a contingency table, given the marginal totals and some supplementary data, J. Roy. Statist. Soc. Ser. A 124(3), 412-420 (1961).

GARSIDE, R. F., Tables for ascertaining whether differences between percentages are statistically significant at the 1% level, British Med. J 1, 874-876 (1961).

GREGORY, G., Contingency tables with a dependent classification, Austral. J. Statist. 3(2), 42-47 (1961).

GRIZZLE, J. E., A new method of testing hypotheses and estimating parameters for the logistic model, Biometrics 17(3), 372-385 (1961).

KENDALL, M. G., and A. STUART, The Advanced Theory of Statistics, 2, Charles Griffin and Company, London, 1961.

OKAMATO, M., and G. ISHII, Test of independence in intraclass 2 x 2 tables, Biometrika 48, 181-190 (1961).

ROGOT, E., A note on measurement errors and detecting real differences, J. Amer. Statist. Assoc. 56, 314-319 (1961).

SCHULL, W. J., Some problems of analysis of multi-factor tables, Bull. Inst. Internat. Statist. 28(3), 259-270 (1961).

YATES, F., Marginal percentages in multiway tables of quantal data with disproportionate frequencies, Biometrics 17(1), 1-9 (1961).

1960

BENNETT, B. M., and P. HSU, On the power function of the exact test for the 2 x 2 contingency table, Biometrika 47, 393-398 (1960).

GRIDGEMAN, N. T., Card-matching experiments: a conspectus of theory, J. Roy. Statist. Soc. Ser. A 123(1), 45-49 (1960).

ISHII, G., Intraclass contingency tables, Ann. Inst. Statist. Math. 12, 161-207 (1960); corrections, 279.

KASTENBAUM, M. A., A note on the additive partitioning of chi-square in contingency tables, Biometrics 16(3), 416-422 (1960).

KUPPERMAN, M., On comparing two observed frequency counts, Appl. Statist. 9(1), 37-42 (1960).

LANCASTER, H. O., On tests of independence in several dimensions, J. Austral. Math. Soc. 1, 241-254 (1960), corrigendum 1, 496 (1960).

ROBERTSON, W. H., Programming Fisher's exact method of comparing two percentages, Technometrics 2(1), 103-107 (1960).

SOLOMON, H., Classification procedures based on dichotomous response vectors, (no. 36), in Contributions to Probability and Statistics, Essays in Honor of Harold Hotelling, (I. Olkin et al., eds.) Stanford U. P., Stanford, Cal. 1960, 414-423. (Also in Studies in Item Analysis and Prediction, (H. Solomon, ed.) Stanford U. P., Stanford, Cal. 1961, 177-186.)

YATES, F., Sampling Methods for Censuses and Surveys, 3rd Edition, Griffin, London, 1960.

1959

ANDERSON, R. L., Use of contingency tables in the analysis of consumer preference studies, Biometrics **15**(4), 582-590 (1959).

CHAKRAVARTI, I. M., and C. R. RAO, Tables for some small sample tests of significance for Poisson distributions and 2 x 3 contingency tables, Sankhya **21**(3 and 4), 315-326 (1959).

GOODMAN, L. A., and W. H. KRUSKAL, Measures of association for cross classification II: further discussion and references, J. Amer. Statist. Assoc. **54**, 123-163 (1959).

HALDANE, J. B. S., The analysis of heterogeneity, I, Sankhya **21**(3 and 4), 209-216 (1959).

HOYT, C. J., P. R. KRISHNAIAH, and E. P. TORRANCE, Analysis of complex contingency data, Journal of Experimental Education **27**, 187-194 (1959).

KASTENBAUM, M. A., and D. E. LAMPHIEAR, Calculation of chi-square to test the no three-factor interaction hypothesis, Biometrics **15**(1), 107-115 (1959).

KULLBACK, S., Information Theory and Statistics. John Wiley and Sons, New York, 1959.

KUPPERMAN, M., A rapid significance test for contingency tables, Biometrics **15**(4), 625-628 (1959).

NASS, C. A. G., The chi-square test for small expectations in contingency tables, with special reference to accidents and absenteeism, Biometrika **46**, 365-385 (1959).

SILVEY, S. D., The Lagrangian multiplier test, Ann. Math. Statist. **30**(2), 389-407 (1959).

SOMERS, R. H., The rank analogue of product-moment partial correlation and regression, with application to manifold, ordered contingency tables, Biometrika **46**, 241-246 (1959).

STEYN, H. S., On chi-square tests for contingency tables of negative binomial type, Statistica Neerlandica **13**, 433-444 (1959).

WEINER, I. B., A note on the use of Mood's likelihood ratio test for item analyses involving 2 x 2 tables with small samples, Psychometrika **24**(4), 371-372 (1959).

1958

BLALOCK, H. M., Jr., Probabilistic interpretations for the mean square contingency, J. Amer. Statist. Assoc. **53**, 102-105 (1958).

BROSS, I. D. J., How to use RIDIT analysis, Biometrics **14**(1), 18-38 (1958).

GARSIDE, R. F., Tables for ascertaining whether differences between percentages are statistically significant, British Med. J. **1**, 1459-1461 (1958).

KASTENBAUM, M. A., Estimation of relative frequencies of four sperm types in Drosophila melanogaster, Biometrics **14**(2), 223-228 (1958).

MITRA, S. K., On the limiting power function of the frequency chi-square test, Ann. Math. Statist. **29**, 1221-1233 (1958).

SNEDECOR, G. W., Chi-square of Bartlett, Mood and Lancaster in a 2^3 contingency table, Biometrics **14**(4), 560-562 (1958) (Query).

1957

BROSS, I. D. J., and E. L. KASTEN, Rapid analysis of 2 x 2 tables, J. Amer. Statist. Assoc. **52**, 18-28 (1957).

CORSTEN, L. C. A., Partition of experimental vectors connected with multinomial distributions, Biometrics **13**(4), 451-484 (1957).

EDWARDS, J. H., A note on the practical interpretation of 2 x 2 tables, Brit. J. Prev. Soc. Med. **11**, 73-78 (1957).

LANCASTER, H. O., Some properties of the bivariate normal distribution considered in the form of a contingency table, Biometrika **44**, 289-292 (1957).

MOTE, V. L., An investigation of the effect of misclassification of the chi-square tests in the analysis of categorical data. Unpublished Ph.D. dissertation, North Carolina State College, Raleigh, North Carolina (also Institute of Statistics Mimeo Series No. 182), 1957.

ROY, S. N., Some Aspects of Multivariate Analysis, John Wiley and Sons, New York, 1957.

SAKODA, J. M., and B. H. COHEN, Exact probabilities for contingency tables using binomial coefficients, Psychometrika **22**(1), 83-86 (1957).

WOOLF, B., The log likelihood ratio test (the G-test), Methods and tables for tests of heterogeneity in contingency tables, Annals of Human Genetics **21**, 397-409 (1957).

1956

CORNFIELD, J., A statistical problem arising from retrospective studies, Proc. 3rd Berkeley Symposium **4**, 135-148 (1956).

FISHMAN, J. A., A note on Jenkins' "Improved Method for Tetrachoric r," Psychometrika **20**(3), 305 (1956).

GOOD, I. J., On the estimation of small frequencies in contingency tables, J. Roy. Statist. Soc. Ser. B **18**(1), 113-124 (1956).

GRIDGEMAN, N. T., A tasting experiment, Appl. Statist. **5**(2), 106-112 (1956).

LEANDER, E. K., and D. J. FINNEY, An extension of the use of the chi-square test, Appl. Statist. **5**(2), 132-136 (1956).

MAINLAND, D., L. HERRERA, and M. I. SUTCLIFFE, Statistical tables for use with binomial samples - contingency tests, confidence limits, and sample size estimates, New York University College of Medicine, New York, 1956.

ROY, S. N., and M. A. KASTENBAUM, On the hypothesis of no "interaction" in a multiway contingency table, Ann. Math. Statist. 27(3), 749-757 (1956).

ROY, S. N., and S. K. MITRA, An introduction to some non-parametric generalizations of analysis of variance and multivariate analysis, Biometrika 43(3 and 4), 361-376 (1956).

WATSON, G. S., Missing and "mixed-up" frequencies in contingency tables, Biometrics 12(1), 47-50 (1956).

1955

ARMITAGE, P., Tests for linear trends in proportions and frequencies, Biometrics 11(3), 375-386 (1955).

ARMSEN, P., Tables for significance tests of 2 x 2 contingency tables, Biometrika 42, 494-505 (1955).

COCHRAN, W. G., A test of a linear function of the deviations between observed and expected numbers, J. Amer. Statist. Assoc. 50, 377-397 (1955).

HALDANE, J. B. S., Substitutes for chi-square, Biometrika 42, 265-266 (1955).

HALDANE, J. B. S., A problem in the significance of small numbers, Biometrika 42, 266-267 (1955).

HALDANE, J. B. S., The rapid calculation of chi-square as a test of homogeneity from a 2 x n table, Biometrika 42, 519-520 (1955).

HILL, B., Principles of Medical Statistics, Oxford University Press, 1955.

JENKINS, W. L., An improved method for tetrachoric r, Psychometrika 20(3), 253-258 (1955).

KASTENBAUM, M. A., Analysis of data in multiway contingency tables, Unpublished doctoral dissertation, North Carolina State College, October 1955.

LESLIE, P. H., A simple method of calculating the exact probability in 2 x 2 contingency tables with small marginal totals, Biometrika 42, 522-523 (1955).

MITRA, S. K., Contributions to the statistical analysis of categorical data, North Carolina Institute of Statistics Mimeograph Series No. 142, December 1955.

ROY, S. N., and M. A. KASTENBAUM, A generalization of analysis of variance and multivariate analysis to data based on frequencies in qualitative categorical or class intervals, North Carolina Institute of Statistics Mimeograph Series No. 131, June 1955.

ROY, S. N., and S. K. MITRA, An introduction to some non-parametric generalizations of analysis of variance and multivariate analysis, North Carolina Institute of Statistics Mimeograph Series No. 139, November 1955.

SEKAR, C. C., S. P. AGARIVALA, and P. N. CHAKRABORTY, On the power function of a test of significance for the difference between two proportions, Sankhya **15**(4), 381-390 (1955).

STUART, A., A test of homogeneity of the marginal distributions in a two-way classification, Biometrika **42**, 412-416 (1955).

WOOLF, B., On estimating the relation between blood group and disease, Annals of Human Genetics **19**, 251-253 (1955).

YATES, F., A note on the application of the combination of probabilities test to a set of 2 x 2 tables, Biometrika **42**, 401-411 (1955).

YATES, F., The use of transformations and maximum likelihood in the analysis of quantal experiments involving two treatments, Biometrika **42**, 382-403 (1955).

1954

BROSS, I. D. J., Misclassification in 2 x 2 tables, Biometrics **10**(4), 478-486 (1954).

COCHRAN, W. G., Some methods for strengthening the common chi-square tests, Biometrics **10**(4), 417-451 (1954).

DAWSON, R. B., A simplified expression for the variance of the chi-square function on a contingency table, Biometrika **41**, 280 (1954).

DORN, H. F., The relationship of cancer of the lung and the use of tobacco, American Statistician 8, 7-13 (1954).

GOODMAN, L. A., and W. H. KRUSKAL, Measures of association for cross classification, J. Amer. Statist. Assoc. 49, 732-764 (1954).

KIMBALL, A. W., Short-cut formulas for the exact partition of chi-square in contingency tables, Biometrics 10(4), 452-458 (1954).

MCGILL, W. J., Multivariate information transmission, Psychometrika 19(2), 97-116 (1954).

1952

COCHRAN, W. G., The chi-square test of goodness of fit, Ann. Math. Statist. 23(3), 315-345 (1952).

DYKE, G. V., and H. D. PATTERSON, Analysis of factorial arrangements when the data are proportions, Biometrics 8, 1-12 (1952).

1951

FREEMAN, G. H., and J. H. HALTON, Note on the exact treatment of contingency, goodness-of-fit and other problems of significance, Biometrika 38, 141-149 (1951).

LANCASTER, H. O., Complex contingency tables treated by the partition of chi-square, J. Roy. Statist. Soc. Ser. B 13, 242-249 (1951).

SIMPSON, C. H., The interpretation of interaction in contingency tables, J. Roy. Statist. Soc. Ser. B 13, 238-241 (1951).

1950

TOCHER, K. D., Extension of the Neyman-Pearson theory of tests to discontinuous variates, Biometrika 37, 130-144 (1950).

HSU, P. L., The limiting distributions of functions of sample means and application to testing hypotheses, Proceedings of the Berkeley Symposium on Mathematical Statistics and Probability (1945, 1946), University of California Press, Berkeley and Los Angeles.

IRWIN, J. O., A note on the subdivision of chi-square into components, Biometrika 36, 130-134 (1949).

LANCASTER, H. O., The derivation and partition of chi-square in certain discrete distributions, Biometrika 36, 117-129 (1949).

LANCASTER, H. O., The combination of probabilities arising from data in discrete distributions, Biometrika 36, 370-382 (1949), Corrig. 37, 452 (1950).

1948

FINNEY, D. J., The Fisher-Yates test of significance in 2 x 2 contingency tables, Biometrika 35, 145-156 (1948).

SWINEFORD, F., A table for estimating the significance of the difference between correlated percentages, Psychometrika 13, 23-25 (1948).

WINSOR, C. P., Factorial analysis of a multiple dichotomy, Human Biology 20, 195-204 (1948).

YATES, F., The analysis of contingency tables with groupings based on quantitative characters, Biometrika 35, 176-181 (1948).

1947

BARNARD, G. A., Significance tests for 2 x 2 tables, Biometrika 34, 123-138 (1947).

BARNARD, G. A., 2 x 2 tables, A note on E. S. Pearson's paper, Biometrika 34, 168-169 (1947).

LOMBARD, H. L., and C. R. DOERING, Treatment of the fourfold table by partial association and partial correlation as it relates to public health problems, Biometrics 3, 123-128 (1947).

PEARSON, E. S., The choice of statistical tests illustrated on the interpretation of data classed in a 2 x 2 table, Biometrika 34, 139-167 (1947).

1946

CRAMER, H., Mathematical Methods of Statistics, Princeton University Press, 1946.

1945

NORTON, H. W., Calculation of chi-square for complex contingency tables, J. Amer. Statist. Assoc. 40, 251-258 (1945).

1943

WALD, A., Tests of statistical hypotheses concerning several parameters when the number of observations is large, Transactions Amer. Math. Soc. 54, 426-482 (1943).

1939

HALDANE, J. B. S., Note on the preceding analysis of Mendelian segregations, Biometrika 31, 67-71 (1939).

ROBERTS, E., W. M. DAWSON, and M. MADDEN, Observed and theoretical ratios in Mendelian inheritance, Biometrika 31, 56-66 (1939).

1938

FISHER, R. A., and F. YATES, Statistical Tables for Biological, Agricultural and Medical Research, Oliver and Boyd, Edinburgh, 1938; 6th ed., 1963 x + 146.

SWAROOP, S., Tables of the exact values of probabilities for testing the significance of differences between proportions based on pairs of small samples, Sankhya 4, 73-84 (1938).

1937

HALDANE, J. B. S., The exact value of the moments of the distribution of chi-square used as a test of goodness-of-fit, when expectations are small, Biometrika 29, 133-143 (1937).

1935

BARTLETT, M. S., Contingency table interactions, J. Roy. Statist. Soc. Supplement 2, 248-252 (1935).

IRWIN, J. O., Tests of significance for differences between percentages based on small numbers, Metron 12(2), 83-94 (1935).

WILKS, S. S., The likelihood test of independence in contingency tables, Ann. Math. Statist. 6, 190-196 (1935).

1934

FISHER, R. A., Statistical Methods for Research Workers, 1934, Oliver and Boyd, Edinburgh, 5th and subsequent editions, Section 21.02.

YATES, F., Contingency tables involving small numbers and the chi-square test, J. Roy. Statist. Soc. Supplement 1, 217-235 (1934).

1924

FISHER, R. A., The conditions under which chi-square measures the discrepancy between observation and hypothesis, J. Roy. Statist. Soc. 87, 442-450 (1924).

1922

FISHER, R. A., On the interpretation of chi-square from contingency tables, and the calculation of P, J. Roy. Statist. Soc. 85, 87-94 (1922).

1916

PEARSON, K., On the general theory of multiple contingency with special reference to partial contingency, Biometrika 11, 145-158 (1916).

1915

GREENWOOD, M., and G. U. YULE, The statistics of anti-typhoid and anti-cholera inoculations and the interpretation of such statistics in general, Proc. Roy. Soc. Medicine 8, 113-194 (1915).

1912

YULE, G. U., On the methods of measuring association between two attributes, Jour. Roy. Statist. Soc. 75, 579 (1912).

1904

PEARSON, K., Mathematical contributions to the theory of evolution, XIII On the theory of contingency and its relation to association and normal correlation, Draper's Company Research Memoirs, Biometric Series 1, 1904, 35 pp.

1900

PEARSON, K., On the criterion that a given system of deviations from the probable in the case of a correlated system of variables is such that it can be reasonably supposed to have arisen from random sampling, Philos. Mag., [5] **50**, 157-172 (1900).

1898

SHEPPARD, W. F., On the application of the theory of error to cases of normal distribution and normal correlation, Phil. Trans. Roy. Soc. London **A192**, 101-167 (1898).

APPENDIX

In this appendix we consider relations connecting the moment constraints, the natural parameters (taus) and their covariance matrices. These relations form the basis for quadratic approximations and the iterative algorithm for the k-sample Newton-Raphson type procedure discussed in Chapter 5.

Let us consider the space Ω and the discrimination information discussed in Chapter 3 Section I and Chapter 5 Section III. Suppose now, for example, that we have three linearly independent constraints or statistics defined over the space Ω

$$T_1(\omega),\ T_2(\omega),\ T_3(\omega). \tag{A.1}$$

Let us determine the value of $p(\omega)$ which, for $\pi(\omega)$ given, minimizes the discrimination information

$$I(p:\pi) = \Sigma\ p(\omega)\ \ell n\ (p(\omega)/\pi(\omega)) \tag{A.2}$$

over the family of distributions $p(\omega)$ which satisfy the moment constraints

$$\Sigma\, T_1(\omega)p(\omega) = \theta_1^*$$

(A.3)
$$\Sigma\, T_2(\omega)p(\omega) = \theta_2^*$$

$$\Sigma\, T_3(\omega)p(\omega) = \theta_3^*$$

where the moment parameters θ_1^*, θ_2^*, θ_3^* are specified values.

If $\pi(\omega)$ satisfies the constraints (A.3) then of course the minimum value of $I(p:\pi)$ is zero and the minimizing distribution is $p^*(\omega) = \pi(\omega)$. More generally, in accordance with the principle of MDI estimation (see Chapters 3 and 5) the minimizing distribution is given by the exponential

(A.4)
$$p^*(\omega) = (\exp(\tau_1 T_1(\omega) + \tau_2 T_2(\omega) + \tau_3 T_3(\omega))\pi(\omega))/M(\tau_1,\tau_2,\tau_3)$$

where the normalizing factor (for the natural constraint)

(A.5)
$$M(\tau_1,\tau_2,\tau_3) = \Sigma \exp(\tau_1 T_1(\omega) + \tau_2 T_2(\omega) + \tau_3 T_3(\omega))\pi(\omega)$$

and the τ's, the natural parameters of the exponential distribution, are in essence undetermined Lagrange multipliers whose values are defined in terms of the moment parameters θ_1^*, θ_2^*, θ_3^* by

$$\theta_1^* = \partial/\partial\tau_1\ \ln M(\tau_1,\tau_2,\tau_3) = (\Sigma \exp(\tau_1 T_1(\omega) + \tau_2 T_2(\omega) + \tau_3 T_3(\omega))T_1(\omega)\pi(\omega))/M(\tau_1,\tau_2,\tau_3)$$

(A.6)
$$\theta_2^* = \partial/\partial\tau_2\ \ln M(\tau_1,\tau_2,\tau_3) = (\Sigma \exp(\tau_1 T_1(\omega) + \tau_2 T_2(\omega) + \tau_3 T_3(\omega))T_2(\omega)\pi(\omega))/M(\tau_1,\tau_2,\tau_3)$$

$$\theta_3^* = \partial/\partial\tau_3\ \ln M(\tau_1,\tau_2,\tau_3) = (\Sigma \exp(\tau_1 T_1(\omega) + \tau_2 T_2(\omega) + \tau_3 T_3(\omega))T_3(\omega)\pi(\omega))/M(\tau_1,\tau_2,\tau_3).$$

We now state a number of consequences of the preceding. We note first that $p^*(\omega)$ is a member of an exponential family of distributions generated by $\pi(\omega)$ and as such has the desirable statistical properties of members of an exponential family. In particular $p^*(\omega) = \pi(\omega)$ for $\tau_1 = \tau_2 = \tau_3 = 0$. We may also write (A.4) as

$$\text{(A.7)} \quad \ln(p^*(\omega)/\pi(\omega)) = -\ln M(\tau_1,\tau_2,\tau_3) + \tau_1 T_1(\omega) + \tau_2 T_2(\omega) + \tau_3 T_3(\omega) = L + \tau_1 T_1(\omega) + \tau_2 T_2(\omega) + \tau_3 T_3(\omega)$$

with $L = -\ln M(\tau_1,\tau_2,\tau_3)$. The loglinear representation (regression) in (A.7) for $\ln(p^*(\omega)/\pi(\omega))$ with $T_1(\omega)$, $T_2(\omega)$, $T_3(\omega)$ as the explanatory variables and τ_1,τ_2,τ_3 as the regression coefficients plays an important role in the analysis, as we have seen.

We note next that the minimum value of the discrimination information (A.2) is

$$\text{(A.8)} \qquad I(p^*:\pi) = \tau_1\theta_1^* + \tau_2\theta_2^* + \tau_3\theta_3^* - \ln M(\tau_1,\tau_2,\tau_3)$$

where the θ^*'s, the moment parameters, are defined in (A.3) and the natural parameters (taus) are determined to satisfy (A.6). Using the loglinear representation in (A.7) it may be shown that if $p(\omega)$ is any member of the family of distributions satisfying (A.3), then

$$\text{(A.9)} \qquad I(p:\pi) = I(p^*:\pi) + I(p:p^*).$$

The pythagorean property (A.9) plays an important role in the analysis of information tables.

We note thirdly relations connecting the moment parameters, θ^*'s, the natural parameters, τ's, and the covariance matrix of the $T(\omega)$'s. If we define the matrices

$$(\underset{\sim\sim}{d\theta^*})' = (d\theta_1^*, d\theta_2^*, d\theta_3^*), \quad (\underset{\sim\sim}{d\tau})' = (d\tau_1, d\tau_2, d\tau_3)$$

then (Kullback, 1959, 49)

$$\text{(A.10)} \qquad (\underset{\sim\sim}{d\theta^*}) = \underset{\sim}{\Sigma}^*(\underset{\sim\sim}{d\tau}), \quad (\underset{\sim\sim}{d\tau}) = \underset{\sim}{\Sigma}^{*-1}(\underset{\sim\sim}{d\theta^*})$$

where $\underset{\sim}{\Sigma}^*$ is the covariance matrix of $T_1(\omega)$, $T_2(\omega)$, $T_3(\omega)$ for the distribution $p^*(\omega)$, that is, with

$$\text{(A.11)}\quad \sigma^*_{ij} = \Sigma\,(T_i(\omega) - \theta^*_i)(T_j(\omega) - \theta^*_j)p^*(\omega),\ \underset{\sim}{\Sigma}^* = (\sigma^*_{ij}),$$

$$\underset{\sim}{\Sigma}^{*-1} = (\sigma^{*ij}), \qquad \partial\theta^*_i/\partial\tau_j = \sigma^*_{ij},\ \partial\tau_i/\partial\theta^*_j = \sigma^{*ij}.$$

From (A.5) it is seen that $M(\tau_1,\tau_2,\tau_3)$ is the moment-generating function of $T_1(\omega)$, $T_2(\omega)$, $T_3(\omega)$ under the distribution $\pi(\omega)$, hence the cumulant-generating function is given up to quadratic terms by

$$\text{(A.12)}\quad \ln M(\tau_1,\tau_2,\tau_3) \approx \theta_1\tau_1 + \theta_2\tau_2 + \theta_3\tau_3 + 1/2 \sum_{ij}\Sigma\, \sigma_{ij}\,\tau_i\tau_j$$

where

$$\text{(A.13)}\quad \theta_1 = \Sigma\, T_i(\omega)\pi(\omega),\ \sigma_{ij} = \Sigma\,(T_i(\omega) - \theta_i)(T_j(\omega) - \theta_j)\pi(\omega).$$

Thus, using (A.12) in (A.6), we get (see equation (26) in Chapter 5 Section IIIB)

$$\text{(A.14)}\quad \begin{aligned} \theta^*_1 &\approx \theta_1 + \sum_j \sigma_{1j}\tau_j \\ \theta^*_2 &\approx \theta_2 + \sum_j \sigma_{2j}\tau_j \\ \theta^*_3 &\approx \theta_3 + \sum_j \sigma_{3j}\tau_j \end{aligned}$$

and then using (A.14) in (A.8) yields (see Chapter 3 Section VIII)

$$\text{(A.15)}\quad 2I(p^*:\pi) \approx (\underset{\sim}{\theta}^* - \underset{\sim}{\theta})'\Sigma^{-1}(\underset{\sim}{\theta}^* - \underset{\sim}{\theta}) \approx \underset{\sim}{\tau}'\underset{\sim}{\Sigma}\,\underset{\sim}{\tau}.$$

We have used three statistics $T_1(\omega)$, $T_2(\omega)$, $T_3(\omega)$ (three moment constraints) thus far in the discussion merely as a matter of convenience. We note that (A.15) holds for a set of m statistics $T_i(\omega)$, $i = 1, \ldots, m$ with appropriate meanings for the matrices. Let us partition the set of m statistics $T_i(\omega)$ into a set H_a say of m_a and a set H_c of the remaining $m_c = m - m_a$ statistics, where the statistics in the set H_a have the property that

(A.16) $$\theta_i^* = \theta_i, \ i = 1, \ldots, m_a.$$

We have the related partitioning of the covariance matrix of the $T_i(\omega)$, $i = 1, \ldots, m$

(A.17) $$\underset{\sim}{\Sigma} = \begin{bmatrix} \underset{\sim}{\Sigma}_{aa} & \underset{\sim}{\Sigma}_{ac} \\ \underset{\sim}{\Sigma}_{ca} & \underset{\sim}{\Sigma}_{cc} \end{bmatrix}, \quad \underset{\sim}{\Sigma}_{ca} = \underset{\sim}{\Sigma}'_{ac}$$

and the $\underset{\sim}{\theta}$, $\underset{\sim}{\theta}^*$, and $\underset{\sim}{\tau}$ matrices

(A.18) $$\underset{\sim}{\theta}^{*\prime} = (\underset{\sim}{\theta}_a^{*\prime}, \underset{\sim}{\theta}_c^{*\prime}), \quad \underset{\sim}{\theta}' = (\underset{\sim}{\theta}'_a, \underset{\sim}{\theta}'_c), \quad \underset{\sim}{\tau}' = (\underset{\sim}{\tau}'_a, \underset{\sim}{\tau}'_c).$$

In terms of the partitionings in (A.17) and (A.18) the relations in (A.14) may be written as

(A.19) $$\begin{aligned} \underset{\sim}{\theta}_a^* &\approx \underset{\sim}{\theta}_a + \underset{\sim}{\Sigma}_{aa}\underset{\sim}{\tau}_a + \underset{\sim}{\Sigma}_{ac}\underset{\sim}{\tau}_c \\ \underset{\sim}{\theta}_c^* &\approx \underset{\sim}{\theta}_c + \underset{\sim}{\Sigma}_{ca}\underset{\sim}{\tau}_a + \underset{\sim}{\Sigma}_{cc}\underset{\sim}{\tau}_c \end{aligned}.$$

It is found that using the fact that $\underset{\sim}{\theta}_a^* = \underset{\sim}{\theta}_a$ and (A.19) in (A.8) now yields (see Chapter 3 Section VIII)

(A.20) $$2I(p^*:\pi) \approx (\underset{\sim}{\theta}_c^* - \underset{\sim}{\theta}_c)'\underset{\sim}{\Sigma}_{cc\cdot a}^{-1}(\underset{\sim}{\theta}_c^* - \underset{\sim}{\theta}_c) \approx \underset{\sim}{\tau}'_c\underset{\sim}{\Sigma}_{cc\cdot a}\underset{\sim}{\tau}_c$$

where $\underset{\sim}{\Sigma}_{cc\cdot a} = \underset{\sim}{\Sigma}_{cc} - \underset{\sim}{\Sigma}_{ca}\underset{\sim}{\Sigma}_{aa}^{-1}\underset{\sim}{\Sigma}_{ac}$ is an $m_c \times m_c$ matrix. The results under the partitioning will help in relating the analysis of information values to those occurring in the testing of subhypotheses in the linear and multivariate linear hypothesis theory (Kullback, 1959, 216 and 259).

We note from (A.6) and (A.7) that

(A.21) $$\begin{aligned} \partial/\partial\tau_i \, \ell n \, p^*(\omega) &= T_i(\omega) - \partial/\partial\tau_i \, \ell n \, M(\tau_1, \tau_2, \ldots) \\ &= T_i(\omega) - \theta_i^*, \end{aligned}$$

hence $T_i(\omega)$ is the maximum likelihood estimator of θ_i^*. Thus if we write $T_i(\omega) = \hat{\theta}_i^*$ and denote the values satisfying (A.6) or (A.14) with $\hat{\theta}_i^*$ in place of θ_i^* and $\hat{\tau}_i$ in place of τ_i, we have corresponding to (A.15)

(A.22) $$2I(\hat{p}^*:\pi) = 2\sum_i \hat{\tau}_i\hat{\theta}_i^* - 2\ \ell n\ M(\hat{\tau}_1,\hat{\tau}_2,\ \ldots)$$

$$\approx (\hat{\underset{\sim}{\theta}}{}^* - \underset{\sim}{\theta})'\underset{\sim}{S}^{-1}(\hat{\underset{\sim}{\theta}}{}^* - \underset{\sim}{\theta}) \approx \hat{\underset{\sim}{\tau}}'\underset{\sim}{S}\,\hat{\underset{\sim}{\tau}}$$

and corresponding to (A.20)

(A.23) $$2I(\hat{p}^*:\pi) \approx (\hat{\underset{\sim}{\theta}}{}_c^* - \underset{\sim}{\theta}_c)'\underset{\sim}{S}_{cc\cdot a}^{-1}(\hat{\underset{\sim}{\theta}}{}_c^* - \underset{\sim}{\theta}_c) \approx \hat{\underset{\sim}{\tau}}_c'\,\underset{\sim}{S}_{cc\cdot a}\,\hat{\underset{\sim}{\tau}}_c.$$

We remark that the covariance matrix of the $\hat{\tau}$'s is the inverse of the covariance matrix of the $T_i(\omega)$'s.

If the $\hat{\theta}_i^*$ are the averages of n independent observations then we have for the MDI statistics

(A.24) $$2nI(\hat{p}^*:\pi) \approx n(\hat{\underset{\sim}{\theta}}{}^* - \underset{\sim}{\theta})'\underset{\sim}{S}^{-1}(\hat{\underset{\sim}{\theta}}{}^* - \underset{\sim}{\theta}) \approx n\ \hat{\underset{\sim}{\tau}}'\underset{\sim}{S}\,\hat{\underset{\sim}{\tau}}$$

and in the partitioned case

(A.25) $$2nI(\hat{p}^*:\pi) \approx n(\hat{\underset{\sim}{\theta}}{}_c^* - \underset{\sim}{\theta}_c)'S_{cc\cdot a}^{-1}(\hat{\underset{\sim}{\theta}}{}_c^* - \underset{\sim}{\theta}_c) \approx n\ \hat{\underset{\sim}{\tau}}_c'\,\underset{\sim}{S}_{cc\cdot a}\,\hat{\underset{\sim}{\tau}}_c.$$

Under the null hypothesis $2nI(\hat{p}^*:\pi)$ in (A.24) or (A.25) is asymptotically distributed as chi-square respectively with m or m_c degrees of freedom.

We remark that the k-sample iteration discussed in Chapter 5 solves the equations (A.6) for the natural parameters (taus) using a Newton-Raphson type procedure and the linearized approximation (A.14).

INDEX

C

D

R

S

T